MACHINES
BUILDINGS
WEAPONRY
OF BIBLICAL TIMES

MACHINES
BUILDINGS
WEAPONRY
OF BIBLICAL TIMES

MAX SCHWARTZ

Fleming H. Revell Company
Old Tappan, New Jersey

Illustrations by Max Schwartz.

Scripture quotations identified KJV are from the King James Version of the Bible.

Scripture quotations contained herein identified RSV are from the Revised Standard Version of the Bible, Copyrighted © 1946, 1952, 1971 by the Division of Christian Education of the National Council of the Churches of Christ in the United States of America, and are used by permission. All rights reserved.

Scripture quotations identified JFB from *The Holy Scriptures,* a Jewish Family Bible According to the Masoretic Text, published by The Menorah Press, Chicago, Ill., copyright © 1955.

Verses marked TLB are taken from *The Living Bible,* copyright © 1971 by Tyndale House Publishers, Wheaton, Ill. Used by permission.

Scripture texts identified NIV are from the Holy Bible, New International Version. Copyright © 1973, 1978, 1984 International Bible Society. Used by permission of Zondervan Bible Publishers.

Scripture quotations identified NEB are from The New English Bible. Copyright © The Delegates of the Oxford University Press 1961, 1970. Reprinted by permission.

Credit lines continue on next page.

Library of Congress Cataloging-in-Publication Data

Schwartz, Max, date
    Machines, buildings, weaponry of bibical times / Max Schwartz.
       p.   cm.
    Bibliography: p.
    Includes index.
    ISBN 0-8007-1630-2. — ISBN 0-8007-5320-8 (pbk.)
      1. Engineering—History. 2. Engineering—Middle East—History.
    3. Science, Ancient. I. Title.
    TA16.S39 1989
    620'.0093—dc20   89-8459
                                    CIP

Copyright © 1990 by Max Schwartz
Published by the Fleming H. Revell Company
Old Tappan, New Jersey 07675
Printed in the United States of America

The author is grateful to the following for use of their material:

ARCHITECTURAL BOOK PUBLISHING CO.: *For passages from Gershon Canaan, REBUILDING THE LAND OF ISRAEL.*

BIBLICAL ARCHEOLOGICAL REVIEW: *Material from Robert J. Bull, "Caesarea Maritima, The search for Herod's City" (May/June, 1982); Volkman Fritz, "Temple Architecture, What Can Archaeology Tell Us About Solomon's Temple" (July/August, 1987); Ernest-Marie La Perrousaz, "King Solomon's Wall Still Supports the Temple Mount" (May/June, 1987); James D. Muhly, "How Iron Technology Changed the World and Gave the Philistines a Military Edge" (Nov./Dec., 1982); Ehud Netzer, "Jewish Rebels Dig Strategic Tunnel System" (July/August, 1988); David Ussiskkin, "Restoring the Great Gate at Lachish" (March/April, 1988).*

DOUBLEDAY: *Excerpts from THE ANCIENT ENGINEERS by L. Sprague de Camp, copyright © 1963 by L. Sprague de Camp. Used by permission of Doubleday, a division of Bantam Doubleday, Dell Publishing Group, Inc.*

DOVER PUBLICATIONS: *Morris Hicky Morgan, VITRUVIUS—THE TEN BOOKS OF ARCHITECTURE.*

FABER & FABER LIMITED: *W.H.G. Armytage, A SOCIAL HISTORY OF ENGINEERING.*

HARPER & ROW, PUBLISHERS, INC.: *Excerpt, adaptations from ARCHAEOLOGY OF THE BIBLE: BOOK BY BOOK, by Gaalyah Cornfeld. Copyright © 1976 by Gaalyah Cornfeld. Reprinted by permission of Harper & Row, Publishers, Inc.*

JOHN JOHNSON LTD.: *Henry Hodges, TECHNOLOGY IN THE ANCIENT WORLD.*

MCGRAW-HILL BOOK CO.: *Harold E. Babbitt, WATER SUPPLY ENGINEERING, © 1939. W. E. Rossnagel, HANDBOOK OF RIGGING, © 1964.*

MIT PRESS: *Bertrand Gille, ENGINEERS OF THE RENAISSANCE.*

MOONRAKER PUBLICATION: *Richard A. Mansir, A MODELER'S GUIDE TO ANCIENT AND MEDIEVAL SHIPS.*

READER'S DIGEST ASSOC.: *Text from ATLAS OF THE BIBLE, copyright © 1981 The Reader's Digest Association, Inc. Used by permission. Text from GREAT PEOPLE OF THE BIBLE AND HOW THEY LIVED, copyright © 1974 The Reader's Digest Association, Inc. Used by permission.*

UNIVERSITY OF CALIFORNIA PRESS: *J. G. Landels, ENGINEERING IN THE ANCIENT WORLD.*

JOHN WILEY & SONS, INC.: *John A. Bateman, INTRODUCTION TO HIGHWAY ENGINEERING, © 1928, 1934, 1939, 1942 by John H. Bateman. Curtis Maitland Brown, EVIDENCE AND PROCEDURES OF BOUNDARY LOCATION, © 1962 by John Wiley & Sons, Inc. Thaddeus Merriman, AMERICAN CIVIL ENGINEER'S HANDBOOK, © 1911, 1912, 1916, 1920, 1930, 1947 by John Wiley & Sons, Inc.*

Contents

Introduction

On the stage of the ancient Holy Land the early civilizations played out the drama of history as they came to power, ruled, and disappeared. The Bible records many of the events of these powerful empires—but we may not recognize the behind-the-scenes influence dramatic innovations in science, engineering, and warfare had upon this panoply.

All the great powers that conquered the ancient Mideast relied on newly acquired technical knowledge to build their great palaces, fortresses, tombs, water and sanitation systems, transportation systems, and weapons of war. These tools helped maintain these societies.

Machines, Buildings, Weaponry of Biblical Times focuses on the history of the ancient science of engineering, which reached new peaks with every successive empire that ruled the Mideast—Egyptian, Assyrian, Babylonian, Persian, Greek, and Roman. Using the Bible as a major source and including evidence from archaeological finds and ancient texts, I have sought to show the interaction between engineering and the biblical world.

By *biblical world* I mean the Mideast and eastern Mediterranean region during the time of the events recorded in the Old and New Testaments. The Bible speaks of engineering as early as Genesis, where God told Noah how to build the ark:

> Make a boat from resinous wood, sealing it with tar; and construct decks and stalls throughout the ship. Make it 450 feet long, 75 feet wide, and 45 feet high. Construct a skylight all the way around the ship, eighteen inches below the roof; and make three decks inside the boat—a bottom, middle, and upper deck—and put a door in the side.
>
> Genesis 6:14–16 TLB

The "tower of Babel," the tabernacle, the walls of Jericho, Hezekiah's water tunnel—the pages of the Bible are full of engineering. This book will consider some of these things, as well as extrabiblical advances in science and engineering during the biblical period. Historians differ on the exact dating, but for our purposes we will consider that the *biblical period* began about 1682 B.C., the time of Abraham, and continued through the end of the Roman period, about A.D. 324.

My writing springs from my lifelong association with architecture and military and civil engineering. Though my first love has always been architectural history, I became a civil engineer after a stint with the United States Army Corps of Engineers and enrollment in the University of Southern California School of Engineering.

But this book also is a cultural and spiritual exploration by a Jew searching for his past. I have sought to relate my educational, spiritual, and professional life experiences to those of my ancient and biblical counterparts. Although I examine the biblical period, I do not try to prove or disprove any aspect of the Bible. I only intend to interpret the past, to shed more light on history, and to provide a clearer understanding of the state of engineering knowledge in the ancient world. In this way, I hope the reader can gain new insight into the Bible.

1
The Ancient Engineer

In biblical times, massive construction projects required complex masonry works, the moving of heavy stones, and some intricate mathematics. Then, as now, engineers needed great skill and knowledge to perform their tasks well.

Our word *engineer* comes from the Latin *ingeniator*, which described those who built engines of war. Throughout history, the military has been responsible for training and employing most engineers—even though these people have worked on many "civil" projects, such as roads, bridges, and harbors. Not until the eighteenth century did engineering become a distinct civilian profession. Even in American history, we find the United States Army Corps of Engineers responsible for many structures and passageways used by the general population. In the same way, most major engineering works of ancient times—the aqueducts, drainage systems, harbors, highways, and public buildings—were government financed, designed by military engineers, and constructed by soldiers or captive laborers.

As in the modern military, when a new weapon of war was introduced, the opposing power invented a counterweapon for defense. For example, the fortress was perhaps the first "war machine," built to defend a settlement from being overrun by its enemies. Initially fortresses had to provide defense against climbing soldiers and flying arrows and spears. So the earliest military engineers built high masonry walls with protective moats and glacis (sloped embankments) outside the walls. These defensive innovations forced attackers to develop tall, wooden, hide-covered siege towers that would reach the top of the walls and protect the soldiers below, who pounded the walls with battering rams. This, in turn, made new defensive machinery necessary.

Roman Engineers

The Romans, civilization's great borrowers, made the most of the knowledge of others. They learned their basic engineering skills from the Etruscans, who had ruled the Italian peninsula for at least three hundred years before them. As the Romans extended their rule throughout the Mediterranean, they gained more engineering knowledge from the Greeks and from Middle Eastern cultures.

They put all this to good use in developing a civilian infrastructure. Once the empire was established and Rome could enjoy the lengthy Pax Romana, its military engineers turned their attention to roads, bridges, and public buildings. These would enhance Rome's control of its occupied territories and put a distinctly Roman stamp on the whole Mediterranean world.

As in other historical periods of great building projects, the *engineer* or *architect* (the two terms in Latin were virtually interchangeable) was a celebrated member of society—someone who had to have expertise in many areas. First-century Roman engineer and philosopher Vitruvius described the kind of person the job required:

> He must be literate and able to express himself clearly.
>
> He must be a skilled draughtsman, and capable of drawing plans, elevations and renderings in perspective sketches.
>
> He must be a mathematician, as well as competent in geometric construction and arithmetic.
>
> He must have an encyclopedic knowledge of mythology and legend, to plan pedimental sculptures or friezes.
>
> He must be familiar with several branches of philosophy.
>
> He must understand the basics of acoustic and musical theory, and must know the rudiments of medicine, as it relates to public health.
>
> He must understand the basics of the law and know the legal precedents for various measures concerning drainage rights, lighting, and must be able to prepare a contract which is clear and unambiguous, and will not rise to litigation later.
>
> Finally, since town sites and encampments are to be oriented without a magnetic compass, he must have enough of a basic knowledge of astronomy to work out the directions from the sun and stars, and calibrate sundials at different latitudes (De Camp 1963, 27).

Not only did the ancient world develop building techniques, it also came up with codes that insured public safety. Laws and other historical documents from Babylon, Greece, and the Roman Empire show that even then engineers had a responsibility for the safety of others.

If a builder build a house for a man and does not make its construction firm and the house which he built collapse and cause the death of the owner of the house—that builder shall be put to death.

If it cause the death of the son of the owner of the house—they should put to death the son of the builder.

If it cause the death of a slave of the owner of the house—he shall give to the owner of the house a slave of equal value.

If it destroy property, he shall restore whatever it destroyed, and because he did not make the house which he built firm and it collapsed, he shall rebuild the house which collapsed at his own expense.

If a builder build a house for a man and do not make its construction meet the requirements and a wall fall in, the builder shall strengthen the wall at his own expense (Miller 1966, 41).

Because construction and maintenance of irrigation canals were vital for ancient Mesopotamia, its legal code includes many rulings on this subject:

The gentleman who opens his wall for irrigation purposes, but did not make his dike strong enough and hence caused a flood and inundated a field adjoining his, shall give grain to the owner of the field on the basis of those adjoining (De Camp 1963, 46).

Hammurabi (1792–1750 B.C.) wrote to his governors about the repair of canals:

Unto Governor Sid-Iddinam say: Thus saith Hammurabi. Thou shalt call out the men who hold lands along banks of the Damanum-Canal that they may clear out Damanum-Canal. Within the present month shall they complete the work . . . (De Camp 1963, 46).

Ancient history includes numerous accounts of catastrophic failure of buildings and other structures. In the Gospels, Jesus even mentions in passing the collapse of a tower in Siloam (Luke 13:4). The Roman historian Tacitus (A.D. 55–120) describes the failure of an amphitheater grandstand.

In another case, Tacitus describes a disastrous collapse of a structure that occurred in A.D. 27, which caused many casualties. According to his description, a builder by the name of Atilius, coming from the poorest part of the city, obtained a building permit to construct a temporary wood-framed amphitheater near the town of Fidenae in the valley of Tiber. At the first exhibition held in the amphitheater, shortly after the crowd had taken their seats, the entire structure collapsed, burying the spectators in debris.

The people of Fidenae were shocked by the accident, which according to the official investigation, accounted for 50,000 casualties, either killed or injured. The town's noblemen provided their palaces as hospitals while aristocratic women served as nurses to the injured.

As in the present day, following a catastrophe of this type the local government or Senate issued new building-code regulations for safety of public buildings used for assembly, and as punishment, the contractor, Atilius, was deported from Italy.

Instead of an actual building code, the early Greeks had a complex collection of requirements for the contractual relationship between builder and client. Restoring fragments of a building dating to 341 B.C., Heinrich Latterman found the names of both the contractor and the building inspector engraved on stone, as well as the date of completion. Certain work specifications were also inscribed.

A typical building specification for the construction of stone masonry walls would describe the manner in which the stones are fitted together, inspection by the architect, insertion of iron dowels into each stone, and the locking of the dowels with molten lead. The specifications also required that the building inspector should be present during the installation of the dowels and lead before the work was covered up.

These ancient specifications are very similar to the regulations of a present-day building code and the construction specifications prepared by modern-day architects and structural engineers.

2
The Land

The land is both canvas and palette for the engineer. On it, he positions his creations—the palaces, temples, aqueducts, and harbors. From it, he also draws his materials. To fully understand the work of ancient engineers, we need to look at the land in which they lived.

Although we will cover advances in engineering from Egypt to Rome, our consideration centers on the Middle East, particularly the land of Israel. Not only has this little strip of earth proven historically significant, it also offers a rich variety of geographical and geological features. We will get a good idea of the engineer's struggle with the land as we look at this region.

Many civilizations have left their mark on the Holy Land. It was known as *Canaan* in Abraham's time, then *Eretz Israel* ("the land of Israel") after the Israelites conquered it. A rebellion following the death of King Solomon divided the land into *Israel* in the north (also known as *Samaria,* for its capital, or *Ephraim,* for its largest tribe) and *Judah* in the south. Under the Romans, the southern region was called *Judea.* They renamed the entire area *Palestina,* a derivation of *Philistina,* the name of the coastal region, where the Philistines had lived. Under Ottoman and British rule, the name changed to *Palestine,* and with the establishment of the Israeli state in this century, it became *Eretz Israel* again.

In its lengthy history, this land has been occupied by Egyptians, Phoenicians, Israelites, Babylonians, Persians, Greeks, Romans, Arabs, Ottoman Turks, and western Europeans. All left their remembrances. Each new society has literally built on the foundations of the old one. Thus many structures uncovered by modern archaeologists are composites of various periods and cultures.

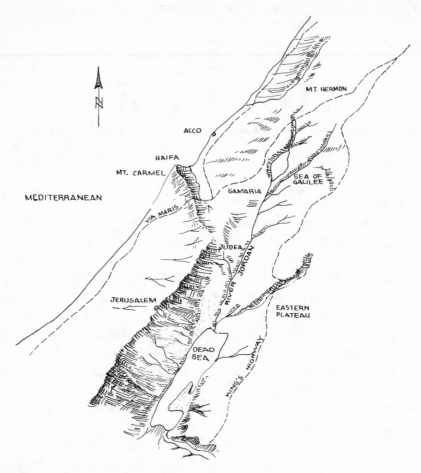

Figure 1: Topographical View of Ancient Israel
The Dead Sea and Jordan River follow the Great Rift Fault, one of the
lowest-elevation regions on earth. The Judean hills rise immediately to the
west of the Dead Sea, creating a ridge that runs northward and then veers
toward the coast. To the west of this ridge are rolling foothills and a
coastal plain.

Along the northern Mediterranean coast lay woodlands. Grasslands
flourished eastward, toward the Sea of Galilee and in the foothills. The
Dead Sea area and southward was very arid, growing only desert shrubs
(Canaan 1954, 74; Carta 1977, 36; Reader's Digest Editors *Atlas* 1981, 39;
Reader's Digest Editors *People* 1981, 315).

Geography

Only 150 miles long and 54 miles across its greatest width, Canaan is about the size of New Jersey. Yet it sports a variety of topographical features and climatic conditions. Its 6,000 square miles include desert, farmland, and mountains.

The land is bounded by the Mediterranean Sea to the west, the Lebanon mountain ranges to the north, the Transjordan desert to the east, and the Gulf of Aqaba on the south. We can denote five major regions of the land itself:

The Galil, or Galilee, rather mountainous (some peaks reach 3,000 feet)

The valleys, including Zevulun (southeast of Haifa), Jezreel, Jordan, and Beth-shean

The coastal plain, from Haifa south to Migdal Gad.

The hill areas, including the hills from Samaria through Judea (where Jerusalem is located)

The Negev, the triangle of desert in the south

An east-west cross section of the land (figure 2) shows the extreme topographic changes. The land rises from sea level, at the Mediterranean, to 2,500 feet above sea level, at Jerusalem, then plunges to 1,250 feet below sea level, at the Dead Sea.

Mountains parallel the coastline, providing a mild, subtropical climate for most of the land. To the east, the Jordan Valley is hot and arid, burned by the wind that sweeps in from the Arabian deserts. But the coastal plain and foothills west of the mountain ridge enjoy moist air from the sea and regular rainfall.

The prevailing winds in the eastern Mediterranean create a natural division of the land into northern, central, and southern zones. The north has abundant rainfall, the center has moderate moisture, and the south is dry.

Intermittent rains generally begin in mid-October, peak in January, and end in the spring. This is when the crops grow. In the summer, the intense heat and arid air burn up the vegetation.

The land also contains many important minerals. Copper, manganese, and barites are found near Elath, and the Dead Sea has iodine, bromine, calcium, and salt. Near the Lebanon border there are clay and basalt, and glass sand is mined at Beersheva. Galilee has pozzolana, Gezer has marble, and bituminous limestone is found near Haifa and Jerusalem. Other mineral resources include sulfur,

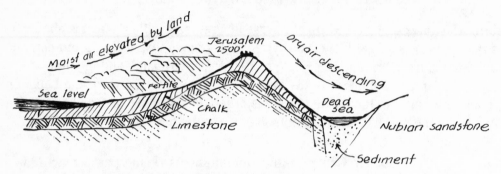

Figure 2: Cross Section of the Holy Land
This cross section of Canaan, cutting across the land at the latitude of
Jerusalem, shows both topographical and climatic features of the region.
Winds from the Mediterranean Sea push moist air eastward, and the rising
terrain pushes it up, where it turns to rain, which waters the western
slopes of the Judean hills. Peaking at Jerusalem, the now dry air descends
into the Jordan Valley.

   Beneath the fertile soil lies granite bedrock covered by limestone and
chalk. East of the Jordan, Nubian sandstone sits atop the granite. The
Great Rift Fault, in which the Jordan River and Dead Sea lie, is lined with
sediment.

gypsum, fire clay, kaolin, feldspar, and potash (Canaan 1954, 74; Reader's Digest
Editors *Atlas* 1981, 38).

   Jerusalem, one of the major cities we will consider, has a rich geological
mixture: various types of colored limestones, much of which can be easily cut
and used in building. Located on the ridge of the Judean mountains, Jerusalem
is bounded on the east by the Judean desert. One ancient caravan route wound
through that desert, from Jericho to Jerusalem and on to the Mediterranean.
Another major route ran north-south along the Judean mountains. Thus Jerusalem lay at a crossroads. Yet its position on the mountains made the city easy
to defend, and a nearby spring supplied its water needs.

   This spring, the Gihon, still gurgles near the present village of Silwan, in the
Kidron Valley. King Hezekiah built an impressive tunnel in 701 B.C. to lead its
waters inside the walled area of the city, to the Pool of Siloam (2 Kings 20:20).
Both Old and New Testaments mention Siloam (John 9:7; Nehemiah 3:15).

   Most of the existing houses of Silwan are built of stone block, as they were in biblical times, with some recent concrete-block additions. As in ancient days, the flat, plastered roofs collect the rain, funneling it to a pipe that leads to a water barrel or underground cistern. The houses cluster on both sides of the steep valley, among the ancient caves. A simple footpath winds among the dwellings. A narrow paved road traverses the bottom of the valley, leading toward the spring and reservoir.

3
Ancient Dwelling Places

You can still see the remnants of the history of housing throughout the Mideast. Housing started with caves, such as those in the Silwan village of modern Jerusalem. Nomadic tents were the next step—such as those still used by Bedouins in the Negev and Jordanian deserts. Then our biblical ancestors developed homes of sun-dried and kiln-baked brick or simple stone-and-rubble huts such as those now found in Hebron and Acco (Morgan 1960, 38).

Tent Dwellings

The people of Israel must have enjoyed the nomadic life, since some remained in tents even after towns and villages sprang up.

Nomadic people needed light, portable shelter, made of readily available materials, so they made tents of wood poles and skins or cloth.

A framework of wooden poles cut from trees formed the basis of the tent. Down the center of the structure the tent dweller erected three poles, each about two meters long, which formed the peaked roof. Beside them, to either side, ran a parallel row of slightly shorter poles that formed the dwelling's walls.

Over this stretched a covering of camel skin, goatskin or goat's-hair cloth, sewn together to the proper dimensions. Using the leather thongs that formed loops in each corner, the tent dweller stretched the material over the poles until it created a taut building held by wooden stakes. Neither the elements nor the impact of humans or animals could easily topple it.

At the doorway, folds of fabric protected the opening. The floor was merely beaten earth, often covered with woven rugs. A fabric curtain or reed screen would divide the tent's interior into two rooms. Visitors were welcome in the front section, but the owner reserved the back room for private and family life. A wealthy man might have separate tents for his wives, as Abraham and Jacob did (Genesis 24:67; 31:33).

Large tents were commonly used for public meetings. The Old Testament, for instance, gives specifications for the elaborate tabernacle used by the wandering Israelites for public worship (Exodus 26, 27, 36–40).

Houses of Clay Brick

When people forsook the nomadic life and settled down, they needed more durable homes. In settlements such as Jericho, the oldest known town in the world, builders made huts with walls of sun-dried clay bricks. Roofs consisted of wooden beams, branches, and grass. In later times, builders covered the roofs with clay, forming a surface strong enough to walk on and smooth enough to channel rainwater down into a waiting jar or cistern.

Before forming clay into bricks, the builder added straw or other fibrous material to the mixture. This provided the tensile strength lacking in the clay itself and kept the bricks from cracking when dried. The Bible assumes the necessity of straw in brickmaking when it tells how the Egyptian taskmasters stopped supplying straw to the captive Israelites, but demanded the same quota and quality of bricks (Exodus 5:7–19). Making bricks without straw was apparently not an option; the Israelites had to gather their own straw and still produce just as many bricks (Morgan 1960, 42).

About two feet thick and seven or eight feet high, walls of sun-dried bricks provided rather good insulation, deflecting the direct heat of the sun but retaining some warmth for the cold nights. In later times, a coating of lime plaster made the walls relatively waterproof.

The roof was held up not only by the clay walls, but also by wooden beams spaced throughout the house's perimeter. The bricks would be laid around the end supports of unfinished tree limbs, which were sometimes hewn into rectangular shapes.

Common people's homes usually consisted of one room adjoining a walled-in yard. The roof doubled as a parlor and sleeping area. Some homes had granaries nearby, either in lined underground pits or above ground, in domed structures made of beaten earth.

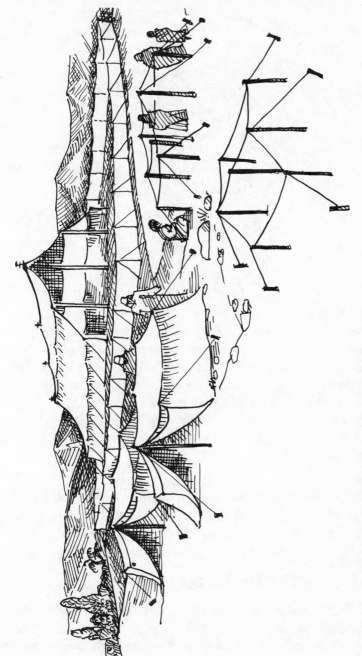

Figure 3: Tents for Family Dwelling and Public Use

In the foreground, to the right, note the arrangement of posts and ropes, ready for the tent covering. Behind them lie completed family tents.

In the background is a representation of the Israelite tabernacle. The courtyard is enclosed by linen sheets stretched out on bronze posts and silver curtain rods. The main tent materials included woven fabric, goatskin, dyed ram's skin, and other light animal hides, supported by posts of acacia wood (Miller 1978, 28, 29).

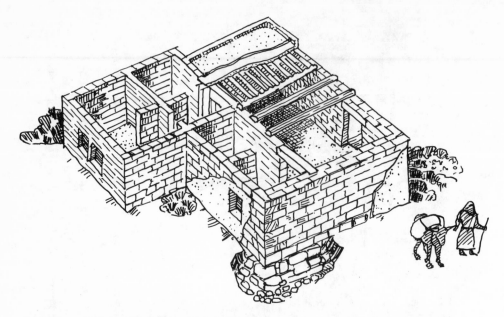

Figure 4: A Typical Clay-Brick House

The cutaway portions in this drawing show the details of the construction of clay-brick houses. The foundation is made of uncut stones and rubble, the walls of clay bricks that have been reinforced with straw, sun dried and coated with a mud plaster. Wooden beams serve as lintels across the tops of doors and windows. Timber beams also form the roof. Covered by branches and twigs, this would support a clay deck finished with lime plaster, sturdy enough to walk on. The floor of the dwelling would be stamped earth (Reader's Digest Editors *People* 1981, 140).

Unfortunately, we have few remains of clay-brick homes from ancient times. Rain, wind, and warfare have chipped away at the clay over the centuries, reducing it back to the soil from whence it came.

Stone Buildings

Since much of the land of Canaan is rich in limestone, this became a widely used building material. Even the common people found it fairly easy to gather uncut stone blocks and pile them up, forming walls. Sometimes they even coated them with a lime plaster. Wealthier homes used carefully cut and dressed stones.

Because of the weight of the stone walls, these homes needed strong foundations, wide enough to distribute the weight over the subsoil, so the building would not sink, and deep enough to keep water from seeping in during the rainy season. Thus builders would dig trenches and lay large stones in them. Because the blocks were not uniform in size, the foundation trenches might be dug at different depths—whatever was necessary to make the tops of these stones level. Where especially wide foundations were needed, two rows of large stones would be laid along the outside edges of the footing, and the space between them would be filled with smaller stones (Paul 1973, 30–33).

In the palace at Tell al-'Ajjul, south of Ashkelon, along the Mediterranean coast, archaeologists have excavated a better-engineered foundation, dating back to 3200–3000 B.C. The trench was quarried out of bedrock, and carefully

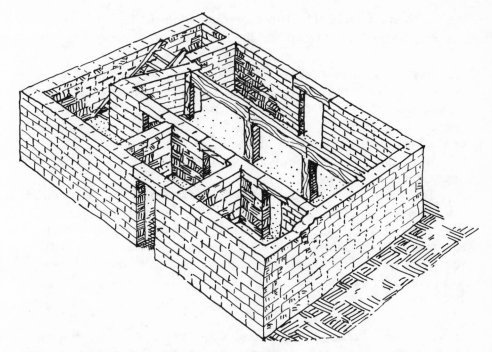

Figure 5: A Typical Stone-Block Dwelling
In this drawing of a house without its roof, we see the details of a home made of stone blocks. Note the wooden posts inside the house and the door lintels of stone. The wooden stairway leads to an opening in the roof. As with clay-brick houses, the roof would be made of wooden beams and branches, supporting a deck of plastered clay.

cut stones were lined up in it. The upper half of these stones protruded up above the floor line of the house, creating a curb at the base of the masonry walls.

The roofs of typical stone-block houses were the same as those of the clay-brick dwellings. Timber beams and branches supported a layer of clay, covered by lime plaster. Some owners kept cylindrical limestone rollers on their roofs to roll down and compact the clay after each rain (Paul 1973, 30, 31; Reader's Digest Editors *People* 1981, 98).

For the common people, roofing timbers were generally sycamore, though more expensive homes used cedar or cypress. If the building was too wide for a single timber to span, builders might install interior posts of stone or timber. These would support interior beams, which the shorter timbers could reach.

Roman engineer-architect Vitruvius, writing in the first century B.C., advised builders on the selection of wood and the construction of roof decks:

> As soon as common oak boards get damp, they warp and cause cracks in the floors. But if there is no winter oak, and necessity drives, . . . use common oak boards cut pretty thin; for the less thick they are, the more easily they can be held in place by being nailed on. . . . As for Turkey oak or beech or ash, none of them can last a great age.
>
> When the wooden planking is finished, cover it with fern, . . . [or] straw, to protect the wood from . . . the lime.
>
> Then, upon this lay the bedding, composed of stones not smaller than can fill the hand. . . . Mix the broken stones in the proportions, if it is new, of three parts to one of lime; if it is old material used again, five parts may answer to two in the mixture. Next, lay the mixture of broken stone, bring on your gangs, and beat it again and again with wooden beetles into a solid mass, and let it be not less than three quarters of a foot in thickness . . . (Morgan 1960, 202).

As in clay-brick homes, the roof was more than a covering for the house. It became an additional living area, often used for sleeping and dining. Women might use it to dry figs, dates, or flax in the hot sun. In the Book of Acts the Apostle Peter prayed on a roof in Joppa and saw a vision there (Act 10:9–11). The roof was reached either by a stone staircase attached to an exterior wall or by a ladder from the inside, which led up through a hatchway in the roof. For safety, a parapet generally surrounded the roof area.

Unlike clay-brick houses, many remains of stone buildings still exist in Israel, representing a broad range of architectural and decorative styles.

Floors, Doors, and Windows

Common people's homes usually had floors of stamped earth, though some added limestone chips to make it harder. The flooring in houses of the wealthy consisted of stone slabs, often finished with terrazzo, a mixture of cement and marble chips (Morgan 1960, 202).

Vitruvius offered guidelines for flooring:

> ... With concrete flooring ... great pains and the utmost precaution must be taken to ensure its durability. If ... concrete flooring is to be laid level with the ground, let the soil be tested to see whether it is everywhere

Figure 6: Construction of a Clay-Brick Dwelling
Builders are setting wood posts and beams atop newly fashioned walls of sun-dried bricks covered with limestone. On top of the wooden framework they will place branches, to form a deck of clay. Then they will cover it with lime plaster, to waterproof the roof and increase its wearability. The deck will be sloped and curbed so that captured water can be diverted to a cistern.

solid, and if it is, level it off and upon it lay broken stone with its bedding. But if the floor is either wholly or partially filling, it should be rammed down hard with great care (Morgan 1960, 202).

He adds specifications for indoor floors:

On this lay the nucleus, consisting of pounded tile mixed with lime in the proportions of three parts to one, and forming a layer not less than six digits thick. On top of the nucleus, the floor . . . should well and truly laid by rule and level.

After it is laid and set at the proper inclination, let it be rubbed down so that, if it consists of cut slips, the lozenges . . . may not stick up at different levels . . . for the rubbing down will not be properly finished unless all edges are at the same level plane (Morgan 1960, 203).

Doors and windows often received elaborate treatment in wealthier homes. As the stone dwelling developed, architects began to pay more attention to the aesthetic appearance of the facade, balancing shapes, masses, detail, and shadows—and placing the windows and doors accordingly.

Windows were usually closed by shutters. In the Hellenistic period, builders tried a variety of translucent materials as windowpanes: oiled cloth, sheepskin, horn, mica, and gypsum shaved to thin sheets. These became obsolete as clear glass was introduced. Though glass-making goes back to the early Egyptians, and the Phoenicians improved on it, not until Roman times did glass became clear enough to let light through (Paul 1973, 36, 37; De Camp 1963, 178).

Doorways usually consisted of a pair of doorposts anchored to a stone sill or threshold at the bottom and spanned by a wooden or stone lintel across the top. Sometimes the threshold was set a bit higher than the floor, to prevent dirt or water from entering.

A heavy wooden door pivoted on a shaped stone socket embedded in the sill. It generally swung inward. To create wider door frames, the builder could add a center post for support, and two half-doors would pivot at the outside posts and meet at the center. For security, a crossbeam bolted the doors on the inside and was anchored to the masonry wall. When set across the doors, this would effectively lock them against intruders.

The size of the door varied. Doors were generally lower than a man's height in private dwellings, but much higher and wider in temples or palaces.

Figure 7: Royal Residence at Megiddo
This reconstruction, based on ruins found at Megiddo, shows some
features of wealthier homes. The spacious palace is built of cut stone
plastered with lime, on a foundation of uncut stone. It includes an interior
courtyard and a large roof deck, where servants might have dried flax and
milled grain. The stone stairway leads to an observation tower with
ramparts (Canaan, 1954, 5).

4
Public Buildings

We know more about ancient palaces and temples than we do about the homes of the common people. Because these structures were made of durable stone, they have remained, whereas the clay and wood of common dwellings have eroded. Therefore, most existing ancient ruins are public buildings or royal residences—at least their foundations and lower portions of their walls.

Engineers of ancient times also paid a great deal of attention to the tombs of royalty. In some cultures, tombs resembled small palaces, with arcades and porticos, vaulted and domed ceilings, and ornate columns at the entrances, spanned by intricate friezes. While their inhabitants have long since passed away, many of these sturdy structures have stood the test of time, giving us important glimpses into the ancient world.

Development of Architectural Design

Ancient civilizations were not as isolated as we might think. Caravans crossed the desert, and merchant ships sailed the Mediterranean, carrying not only goods, but knowledge, art, and ideas. Different civilizations throughout the Mediterranean world developed their particular architectural styles, but they also borrowed from one another.

Three distinct architectural styles emerged first in the Bronze Age: Western Asiatic, Minoan, and Egyptian.

Out of the Western Asiatic design came the Sumerian, Babylonian, and Hittite architectures, which in turn influenced the style of the Persians, Seleucids, and Sassanians.

From the Greek style came Minoan architecture. Roman techniques incorporated Sumerian, Babylonian, Hittite, and Egyptian forms, though it was most strongly influenced by the Greek style.

From Roman and Persian styles came the Eastern Roman Empire, Byzantine, and Hellenic designs. All these combined to form the Islamic style (Allsopp 1965, 9).

Influences of Architectural Styles

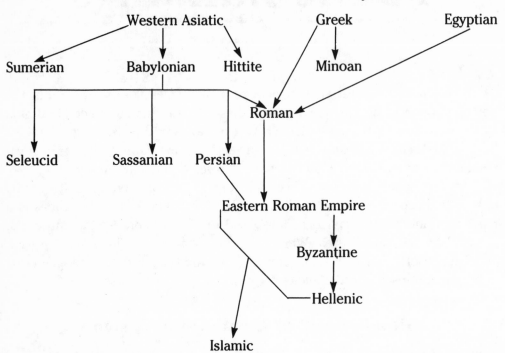

The Hebrews were influenced by many different cultures: Egyptian, Phoenician, Babylonian, Persian, Greek, and Roman. We see the merging styles in the tombs of Absalom and Zachariah, which contain the Egyptian cavetto cornice and pyramidal roof, but Doric and Ionic engaged orders (Hamlin 1904, 35).

Columns

According to Vitruvius, the Doric column (figure 9) was based on the proportions of the human male figure. Its average height is seven times the thickness of the column. The column tapers as it rises, and its fluted shaft contains sixteen to twenty carved channels. Typically the shaft was composed of several shorter

Figure 8: Wall Construction
Builders would dig down to bedrock and there begin assembling a
foundation of uncut stone. On top of this they would place the main wall,
of cut stone (Morgan 1960, 51).

stone cylinders, placed one on top of another. Each piece had to be light enough
for two to four workers to carry and place. The structure above the capital is
known as the entablature, and is divided into the architrave (just above the
column), the frieze and horizontal cornice (the next level, which might include
carved words, pictures, or decorative designs), and the typanum or pediment
(the upper sloping portion, which includes a raked cornice molding) (Landels
1978, 9).

The Ionic column (figures 10–12), Vitruvius said, was based on the female
figure. By extending the height to about eight times the thickness of the column,
its designers produced a more slender appearance. Also, the capital volutes,
scrolling at the top of the column, suggest a woman's hair; the fluted shaft
resembles the folds of a woman's garment; and the base of the column might be
the twisted cords of her sandals. Note also the entablature in this example: the

three flat bands of the architrave and the ornamentation of eggs-and-darts and dentils at the top (Hamlin 1904, 48; Miller 1966, 51).

Some early columns were designed to represent trees and plants. This may reflect the early use of timber to support buildings; now stone was made to copy the material it replaced. The Egyptian date-palm column (figure 13) mimics the palm trunk or bundles of reeds or saplings lashed together. The capitals had appropriate foliage carved in stone. The Egyptian lotus column (figure 14) and clustered lotus column (figure 15) also mirrored the plant world. The Greeks did the same thing with their Corinthian column (figure 16), employing carved acanthus leaves in the capital (Hamlin 1904, 48).

Masonry Arches

The introduction of masonry arches was a major breakthrough in architecture—particularly in the design of public buildings. Previously, the size of the entrance to a room had been limited by the effective span of wood beams. But now arches could support much larger entrances, and by using them for interior design, builders could design much larger rooms.

The engineering involved is fairly simple. An arch consists of a series of shaped stone blocks that gradually span an opening by transferring their weight laterally to the adjacent blocks and eventually to the supporting abutments or walls. The arch essentially creates a beam that has been curved in the plane of the loads. Therefore, any section in an arch may be subjected to the same bending and shear stresses as an ordinary beam. In addition, it is subject to thrust from the components of vertical loads in the direction of the axis of the arch. In simpler terms, each block leans on the block beneath it and, in a way, presses on the block above it.

To ensure pure arch action, a masonry arch must meet three conditions:

1. The length of the span, or distance between supports, must remain constant.
2. The elevation of the ends must remain unchanged.
3. The inclination of the skewbacks (the first stones on the support, which begin the angle of the arch) must be fixed.

If any of these conditions are violated by settlement, sliding, or rotation of the abutment, the arch may fail.

COLUMNS

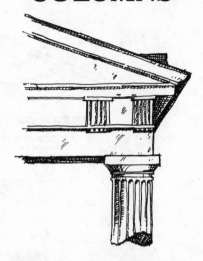

Figure 9: A Greek Doric Column

Figure 10: Front View of a Greek Ionic Column

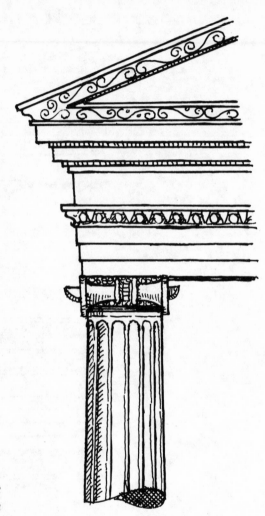

*Figure 11: Side View of the Greek
                Ionic Column*

Figure 12: Detail of the Ionic Scroll

Figure 13: An Egyptian Date-Palm Column

Figure 14: An Egyptian Lotus Column

Figure 15: An Egyptian Column With Clustered Lotus

Figure 16: A Greek Corinthian Column

It is apparent that the engineers and stonemasons of biblical times knew these rules as they built their arches. Just like modern engineers, they probably used the line-of-thrust or graphic analysis method to calculate the arch's dimensions. In fact, the criteria for arch design have changed little from ancient times to the present day (Hamlin 1904, 47–51).

Nebuchadnezzar's Palace

The Babylon built by King Nebuchadnezzar has been called the most magnificent city of the ancient world, yet it lasted only a short time. After the king died in 562 B.C., his heirs failed to preserve his empire.

Nebuchadnezzar's Palace, one of the glories of that great city, lies in ruins today. Still, those ruins give us a glimpse of its majesty. The foundational platform of the palace was 700 yards long, 600 yards wide, and 70 feet high. This raised mound provided safety for the inhabitants—entry was only through the guarded gates. It also had some health benefits—palace residents could enjoy the prevailing breeze and freedom from insects.

The walls, piers, and buttresses were made with pale yellow bricks, set in a fine cement. Resembling modern firebricks, these high-quality Babylonian building blocks were a foot square and three or four inches thick. Each was stamped with the name *Nebuchadnezzar*.

The Hanging Gardens of Babylon

Legend has it that Nebuchadnezzar's wife was homesick for the mountainous land of Media. Babylon, built on the flat river plain, seemed too dull. So the king tried to build her an artificial, tree-capped mountain by the banks of the Euphrates. The result: another wonder of the ancient world, the famed Hanging Gardens of Babylon.

About 400 feet square and 75 feet high, the garden structure consisted of many parallel tiers of open arches. This allowed a solid platform to be built over each tier. A great mass of earth covered the topmost tier, and flowers, shrubs, and trees were planted there.

Using an arrangement similar to the Archimedes' screw, architects managed to provide an irrigation system for the gardens. To prevent water damage to the brick masonry, between the brickwork and the earth, builders installed a layer of reeds mixed with bitumen, then a double layer of burned brick cemented with gypsum, and finally a coating of sheet lead.

BUILDING WITH ARCHES

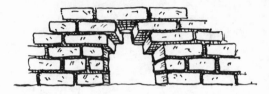

Figure 17: A Stepped Corbeled Arch
This probably represents the earliest form of a masonry arch.

Figure 18: A Straight Stone Arch
This stone post-and-lintel arch at Mycenae is known as the Lion Gate. It consists of dressed, uncoursed masonry, and stone jambs (or posts), at each side of the opening. The posts support a stone lintel, carved into a single column and two rampant lions (Hamlin 1904, 44).

Figure 19: A Square Arch
The columns of the Citadel Gate, built at Hazor during King Ahab's reign, have carved proto-Aeolic capitals and represent construction during the days of the monarchy in Jerusalem (Reader's Digest Editors *Atlas* 1981, 120).

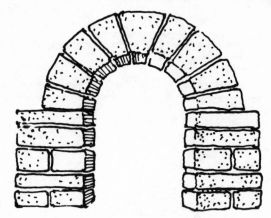

Figure 20: A Simple Round Arch

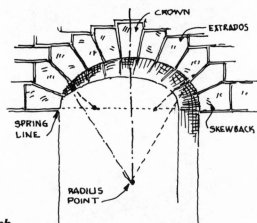

Figure 21: An Elliptical Stone Arch

The main forces of an arch consist of the three radius points, the horizontal spring line, major and minor axes, and rise (the distance from the spring line to the crown). Units of cut stone include the *skewback,* which is at the spring line, the *extrados,* which forms the arch, and the *keystone,* at the crown (Hamlin 1904, 81).

Figure 22: A Pointed Stone Arch
Later known as the Gothic arch, this type contains two radius points, which are at the spring line, with arcs that meet at the crown. The wedge-shaped stones were called *voussoirs* (Canaan 1954, 63).

Figure 23: A Roman Barrel Vault
This structure was formed of semicircular arches and used in the construction of large rooms, culverts, aqueducts, and bridges (Hamlin 1904, 81).

Figure 24: A Groined Vault
This complex structure was called a four-part vault or cross vault.
Examples of this type of construction have been found adjacent to the
western wall at the Temple Mount in Jerusalem. Key elements of this
structure are the spring line, haunch, crown, and keystone (Hamlin 1904,
81).

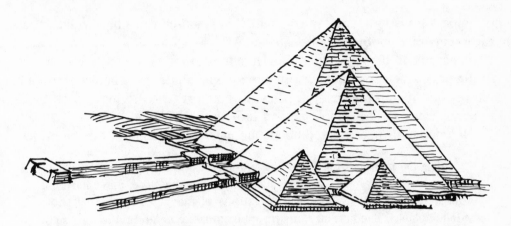

Figure 25: The Great Pyramid of Gizeh
Pyramid building took place between 3000 and 2000 B.C. The Great Pyramid
of the Gizeh group, near Cairo, measured 755 feet at the base and 482 feet
high; it required 2.25 million blocks of stone, with an average weight of 2
tons each. The project required 100,000 workmen and took 20 years to
build (Rossnagel 1961, 1).

Tombs

The Great Pyramid

Built about 2600 B.C., Cheops's pyramid at Gizeh (figure 25) is still the largest
cut-stone structure in the world. Covering an area of thirteen acres (enough for
eight football fields), it rises as high as a forty-story skyscraper. Its square base
measures 755 feet wide.

Napoleon reportedly commented that there were enough blocks in the
pyramid to build a ten-foot wall around France. The massive structure consists
of 2.25 million stone blocks, averaging more than two tons apiece (Rossnagel
1964, 1–3); most of these are coarse desert sandstone quarried locally. The
casing of the pyramid, however, consists of closely fitted limestone, quarried
at Tura, about thirty miles south of Gizeh. Some blocks are as heavy as 15 tons.
One interior burial chamber contains nine blocks of granite, each weighing
about 400 tons. These were quarried at Aswan, about 300 miles upstream on the
Nile.

The pyramid was built for King Khufu (*Cheops* in Greek), of Egypt's Fourth
Dynasty. The methods of its construction remain a mystery. How were those

huge blocks transported and positioned? Horse-and-cart machinery did not appear in Egypt for another eight centuries.

According to the Greek historian Herodotus, Khufu used forced labor, conscripting 4,000 men at a time to work three-month shifts. They hauled the stones from the quarry and put them in place. It took forty men to move each block, Herodotus says, and the whole project took twenty years.

Herodotus also speculated about the machinery required for the project:

> The pyramid was built in steps, battlement-wise, as it is called, or according to others, altar-wise. After laying stones for the base, they raised the remaining stones to their places by means of machines formed of short wooden planks. The first machine raised them from the ground to the top of the first step. On this there was another machine, which received the stone upon its arrival, and conveyed it to the second step, whence a third machine advanced it still higher (Miller 1966, 32–39).

But others have suggested that ramps of earth were used to pull the blocks into position on the upper levels of the pyramid. One Egyptian picture from about 2000 B.C. shows 172 men pulling a sixty-ton statue on a sled, without rollers. In this case, milk or some other fatty liquid might have been poured on the ground to lubricate it. The most common theory holds that the pyramid workers pulled the stones on rollers that were set upon wooden rails.

The pyramid shows remarkable precision in surveying. One side of the base lies within 0.1 degree of a true east-west axis. The other sides were measured with such accuracy that there is only a 7.9-inch error in closure of a true square. Each of the 206 courses of stone is set almost perfectly level—the southeast corner is only a half inch higher than the northwest corner. The builders probably used channels of water to maintain a true horizontal plane.

On the north side of the pyramid, 56 feet above ground, a small entrance opens to a low, narrow tunnel 150 feet long. This leads slightly downward to the grand gallery, a hall 26 feet high. At the center of the pyramid, adjoining this hall, is the main burial chamber, 17 feet by 234 feet, with a ceiling 19 feet high. The structure contains two other burial chambers, one underground (in case the king died before the pyramid was finished), and the third (for the queen) placed between the other two. These are reached by tunnels that branch off from the main tunnel.

Whatever method was used, it is obvious that the construction of this pyramid required four elements (Rossnagel 1964, 1–4):

Figure 26: The Tomb of Absalom
Sometime in the first century
before Christ, this monument,
called the Pillar of Absalom or the
Tomb of Jehoshaphat, was built
east of Jerusalem, in a necropolis
in the Kidron Valley. Although
tradition says it is the burial place
of King David's son Absalom, this
building is not from that era
(Canaan 1954, 18, 19).

Precise use of simple devices
Unlimited manpower
The ability to organize and command that manpower
Unlimited time to complete the project

Tombs of Absalom and Zachariah

In the Kidron Valley, south of Jerusalem's Temple Mount, the Pillar of Absalom
(the Tomb of Jehoshaphat) consists of a circular pyramid rising from a cylinder-
shaped base. This rests on a square platform with a wreathed frieze that appears
to be tied around it like a rope. A cylindrical pyramid forms the top of the tomb.

Adjacent to it is a second tomb, the Tomb of Zachariah, which betrays a
mixture of foreign influences. Greek architecture shows in the Ionic columns, cut

Figure 27: The Tomb of Zachariah
Hewn from the side of a rock, this
tomb includes an Egyptian-style
pyramid above a square tower
decorated with Ionic columns
(Canaan 1954, 18, 19).

in reddish granite, that decorate the front. On top of the tomb is an Egyptian-
style pyramid roof, with an Egyptian groove, or cavetto. The cubelike structure,
carved out of a rock, stands on a stepped base, which encloses the tomb area.

Though tradition associates these tombs with Absalom and the prophet
Zachariah, the architectural styles place their construction later: the second or
third century B.C. for Absalom's tomb and possibly as late as the first century A.D.
for Zachariah's (Canaan 1954, 18, 19).

The Herodium

Herod the Great built the Herodium as his burial place. One of many hilltop
fortresses in Palestine, it is the most impressive. The existing hill was heightened
and the fortress built at the summit, with four round lookout towers and three
concentric walls. As with other hilltop citadels in Palestine, the Herodium is
within signaling distance of one other citadel, so that reinforcements could be
summoned in case of invasion (Netzer 1988, 18–33).

5
Places of Worship

Most nations of the ancient world practiced polytheism. Israel, with its devotion to its one true God, was a notable exception. The others worshiped numerous gods, associating them with various elements of creation—the sky, the earth, the crops, the sea, and so on. When a nation conquered another, it often attempted to impose its gods on the weaker nation. The Book of Daniel, for instance, portrays the Babylonians forcing the captive Jews to bow before an image of gold (Daniel 3).

Such forced conversion, as well as more subtle syncretism, created great similarities in how ancient cultures worshiped their deities. Just as they shared architectural ideas, they also shared ideas about gods. We see both of these facts in the remains of ancient worship places.

Polytheistic Places of Worship

Ziggurats

The early Babylonians built ziggurats to reach their gods. *Ziggurat* means "mountaintop," and it refers to the temple towers built on separate terraces near the temples. These were probably used as observatories—for the Babylonians were early experts in astrology. From these vantage points, priests would watch the moon and stars and chart their positions.

Many cities in Mesopotamia had ziggurats, but the best remaining one is at Ur. This 70-foot-high structure measures 200 feet by 150 feet at its base and has three receding terraces. Each brick terrace was covered with soil and planted with trees and shrubs. The bottom terrace was black, the next red. At the top sat a shrine of enameled blue bricks. Side stairs led to a shrine on the lower level, a central stairway to the shrine at the top.

The Tower of Babel, mentioned in the Bible (Genesis 11:1–9), was probably the famous ziggurat at Babylon, built to honor Marduk, the head of the Babylonian pantheon. After several destructions and rebuildings, Nebuchadnezzar II finally completed the structure about 600 B.C. It stood 300 feet high, covered with colorful enamel bricks.

Temple of Bel

We see more Babylonian engineering skill in the Temple of Bel. This eight-story brick structure resembled a pyramid, with each story a bit smaller than the one below. According to Herodotus, the top floor contained a beautiful adorned temple. Worshipers climbed to the top on outside stairs, and seats were placed partway up, so they could rest.

The temple's structural design was rather simple. All walls were vertical, except for the third story, which had a slight backward tilt, but these walls were reinforced by buttresses. This gives further evidence that the ancient Babylonians had a good knowledge of design principles.

Figure 28: A Ziggurat
The ziggurat was the centerpiece of most Mesopotamian cities, built entirely of sun-dried brick reinforced with layers of reed matting (Reader's Digest Editors *People* 1981, 235).

Babylon had numerous chapels, shrines, and temples. It would not be unusual for a Babylonian city to contain fifty temples, fifty shrines to Marduk, and hundreds more shrines to earth gods. Some cities even "zoned" these buildings into a separate walled section known as "the sacred place."

The Babylonians liked height. Enamored with the gods of the heavens, they built them soaring temples. But they also created the illusion of height with recessed walls and platforms of solid brick. These platforms also protected the temples from flood damage.

Philistine Temples

Except for their struggles with the Israelites, recorded at length in the Bible, the Philistines left little mark on history. These ancient warriors inhabited five cities of the Canaanite plains, including the area known as Gaza. In this area, a recent excavation has uncovered an eleventh-century temple, the Temple of Bethshean. Here the Philistines probably worshipped their fertility god, Dagon. The exterior walls were of masonry, but the interior had two rows of wooden columns that supported the roof. The Book of Judges describes the mighty man Samson pushing apart two pillars in a Philistine temple and literally bringing down the house (Judges 16:23–30). If the design of that temple was similar to this one, we can see how that might happen.

Greek Temples

Many Greek temples remain from ancient times. The earliest of these were built as "god houses," dedicated to one god alone. In some cases, only priests were allowed to enter. In some of the major temples, Greek architects avoided the use of mortar by trimming the stone blocks to precise surfaces and using iron clamps anchored in notches by molten lead.

Jewish Places of Worship

The Jews were different from their neighboring nations. They worshiped not a collection of gods, but a single God, and their God defied physical representation. What did God look like? Not a bull, or a fish, or a frog, as other religions might suggest. Yahweh was not a sky god or an earth god or a sea god, but a God of all being. He had made humans in his image but they were not to make images of him.

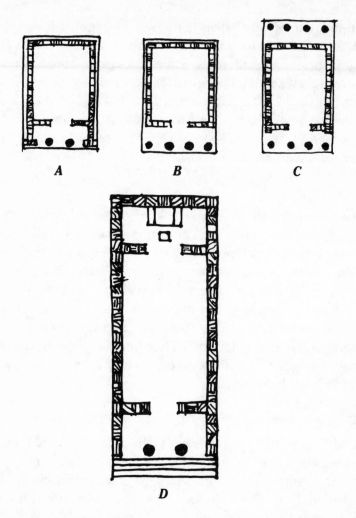

Figure 29: Temple Floor Plans

Designs of temples varied and developed from place to place and through time. Here we see four recurring styles. The temple in antis plan (*A*) has a main hall and a porch whose roof is supported by two columns. The prostyle (*B*) has an open portico in front, with a colonnade of four columns supporting the portico roof. The amphiprostyle (*C*) has porticos in front and back, with rows of columns supporting the roof at both ends. The fourth example (*D*) is the plan of a temple from northern Syria, dating to about 1115 B.C. Similar in style to Solomon's Temple in Jerusalem, it includes a portico (Hebrew, *ulam*), main hall (*hekhal*), and an inner sanctuary (*debir*) (Hamlin 1904, 54).

That stipulation came early in the Israelites' charter, the Ten Commandments: "You shall not make for yourself a graven image, or any likeness of anything that is in heaven above, or that is in the earth beneath, or that is in the water under the earth; you shall not bow down to them or serve them ..." (Exodus 20:4, 5 RSV).

The God of the Israelites could not be simply cooped up in a building. He wrestled with Jacob by a stream (Genesis 32:22–30) and spoke to Moses from a desert bush (Exodus 3:1–6). As the Israelites wandered in the desert, their God was with them. Wherever they set up the tabernacle tent, that became God's holy place.

With the furnishings for the tabernacle, the Israelites began their crafting of sacred artifacts. Skilled artisans were employed to make the objects that would be used in Israel's worship. The law against graven images *of God* remained in effect, but the artistry of the Hebrews could come out in numerous worship symbols. Among those used by builders and craftsmen then and in the following centuries were palm trees (tree of life), lullav (palm branches), ethrog (citron fruit), acanthus leaves, birds, bulls, ram's horn, three crowns, the menorah or seven-branched candlestick (*see* figure 30), Jachin and Boaz (columns), the six-pointed star (the star of David), and signs of the zodiac.

The Ark of the Covenant was the most sacred of the Jews' objects associated with worship. This sacred chest, made along with other tabernacle furnishings, measured about forty-five inches long, twenty-seven inches wide, and twenty-seven inches high. It was overlaid with gold and had four gold rings around it, through which priests could insert poles of acacia wood to carry the Ark. The lid bore two cherubim of hammered gold, with their wings spanning the length of the chest. The Ark became the symbol of God's presence as the Israelites wandered from place to place. Inside it were placed the stone tablets of the Ten Commandments.

Solomon's Temple

When the Jews finally settled in their land, they needed a more permanent place of worship. A single Temple, centrally located, would call all Israel together from their far-flung settlements, to worship the true God. The great King David found the site, Jerusalem. His son Solomon, reigning over what proved to be the economic and military glory years of Israel, built the richly ornamented Temple.

This remains the greatest achievement of Jewish architecture. A variety of

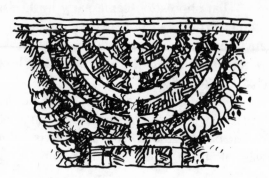

Figure 30: A Carved Hebrew Ornament
This design of the seven-branched lampstand, candelabra or menorah, appears in various forms in early Hebrew buildings (Negev 1980, 104; Reader's Digest Editors *Atlas* 1981, 203).

materials, many of them imported, were used, and Phoenician craftsmen were even brought in to assist on the project. From the biblical descriptions (1 Kings 6, 7; 2 Chronicles 3, 4), it seems to have followed the Egyptian style, with successive courts, lofty entrance pylons, the sanctuary, and the Holy of Holies.

In its ornamentation, the Temple showed workmanship similar to the Phoenician and Assyrian styles, including cedar woodwork, decoration of hammered copper and gold sheets, and metal overlaying.

Built of limestone, the Temple towered 50 feet high and measured 180 feet in length and 90 feet in width. It stood on a sturdy stone platform, with huge retaining walls and impressive vaulted chambers. The main entrance faced east, flanked by two bronze pillars, named Jachin and Boaz. A vestibule led into a large main sanctuary. Banks of small windows, near the ceiling on the north and south walls, illuminated the sanctuary area.

It is said that no stone was visible from the interior of the Temple, since the ceiling was ornamented with sunken-wood panels, and planks of cedar or cypress covered the floor. "He lined the walls of the house on the inside with boards of cedar; from the floor of the house to the rafters of the ceiling, he covered them on the inside with wood; and he covered the floor of the house with boards of cypress" (1 Kings 6:15 RSV).

Note the other details of construction contained in the biblical account:

"He carved all the walls of the house round about with carved figures of cherubim and palm trees and open flowers . . ." (1 Kings 6:29 RSV).

"And he made for the house windows with recessed frames" (1 Kings 6:4 RSV).

". . . And he made the ceiling of the house of beams and planks of cedar" (1 Kings 6:9 RSV).

"For the entrance of the inner sanctuary he made doors of olive-wood. . ." (1 Kings 6:31 RSV).

"In the inner sanctuary he made two cherubim of olivewood, each ten cubits high. . . . And he overlaid the cherubim with gold" (1 Kings 6:23, 28 RSV).

"For the cherubim spread out their wings over the place of the ark, so that the cherubim made a covering above the ark and its poles" (1 Kings 8:7 RSV).

The text also indicates that "neither hammer nor axe nor any tool of iron" was used in the Temple itself, indicating that the stone was precut in the quarry, to specification, and implying that the timbers were fitted with pegs (1 Kings 6:7 RSV).

Remarkably, the Temple was built in seven years, begun by King Solomon's orders in 967 B.C. and completed in 960 B.C. This First Temple stood for nearly four centuries, being destroyed by the armies of Babylonian King Nebuchadnezzar in 587–586 B.C. The Temple had three main sections (*see* figure 31): the unenclosed portico (*ulam*), just 10 cubits deep and 20 cubits wide (a cubit is about a foot and a half); the sanctuary (*hekhal*), measuring 40 cubits by 20 cubits; and the inner sanctuary (*debir*), also called the shrine or the Holy of Holies, 20 cubits by 20 cubits. The height of the vestibule and sanctuary was 30 cubits, but the Holy of Holies had a lower ceiling, forming a perfect 20-cubit cube. Along the side walls and rear walls extended a series of storerooms, two stories high. The exterior walls of the first-story rooms were thicker than those on the second story, to carry its extra weight (Fritz 1987, 38–49; Yaggy 1881, 440; Reader's Digest Editors *Atlas* 1981, 120; Reader's Digest Editors *People* 1981, 188).

The Second Temple

The people of Judea hated King Herod (40–4 B.C.). An Idumean usurper, he had buddied up to the conquering Romans, who became the power behind his throne. As if to prove his devotion to the Jews, Herod authorized massive building projects throughout the land—fortresses, harbors, impressive Greek-style cities. His greatest project was the rebuilding of Solomon's Temple.

It had been over 500 years since the Babylonians had destroyed the First Temple, burning it and stealing its sacred objects. Its glory was a distant memory. The Jews who returned from Babylonian captivity had rebuilt the Temple to a degree in 520–515 B.C., but this, too, had been ravaged and desecrated by conquering powers.

Although Roman engineers and designers were at his disposal, Herod enjoyed acting as his own architect. He based his vision for the Second Temple on Solomon's original, but the new Temple was much more grandiose, incorporating some Greek and Roman elements.

The Jewish historian Josephus details the construction of Herod's Temple, which began in 20 B.C. and was not completed until A.D. 62. Ironically, the masterpiece of Herod and his heirs was enjoyed in its completed form for less than a decade. In A.D. 70, Roman armies quelled a Jewish rebellion and destroyed the Temple once again.

Herod's Temple was built on the same site as Solomon's, at the peak of rocky Mount Moriah. Tradition held that this was where Abraham had nearly sacrificed his son Isaac (Genesis 22:2; 2 Chronicles 3:1). The peak was, of course, leveled off. Herod extended the original platform area, building retaining walls from large ashlar blocks. The resulting Temple Mount area formed a trapezoid about 910 feet wide in the south, 1,550 feet long on the east side, and 1,575 feet long on the west—40 acres, or 172,000 square yards, in all. On the south side, a broad staircase of limestone led up to the Royal Portico and the Court of the Gentiles. The Temple and courtyards stood about 154 feet above the base of the Temple Mount. Within the mount itself were four or five levels of storerooms, stables, and cisterns.

Columned walkways stood around the perimeter of the esplanade. The Royal Portico was the largest, according to Josephus, with 162 monolithic columns, each 4.6 feet in diameter and capped by ornate Corinthian capitals, with rows of acanthus-leaf decorations. Three stories high, this portico included various halls, often used by money changers and merchants who sold ritual objects.

The public would normally have entered from the southern staircase, through the Royal Portico and into the Court of the Gentiles, exchanging any foreign currency for silver shekels and then buying food, drink, or sacrificial animals.

The Court of the Gentiles surrounded the Temple and its interior courts, extending to the north and south. At the northwest corner of the Temple Mount sat the foreboding Antonia Fortress, housing the Roman soldiers assigned to

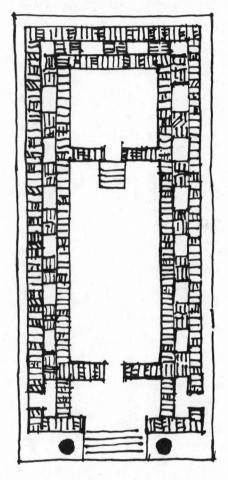

Figure 31: Plan of Solomon's Temple

Jerusalem. This would have been a morose reminder to the Jews of their repression, but Herod might have found it strangely comforting.

On the eastern side of the interior Temple area, a gate (probably the Beautiful Gate, *see* Acts 3:2) led into the Court of Women. The Nicanor Gate led into the narrow Court of Israel, and then the Court of Priests, where the altar stood. Finally, one could enter the Temple itself, crossing through a porch and into the sanctuary.

The roof of Herod's Temple was supported not by pillars but by wooden beams spanning the width of the building, a common element in Middle Eastern architecture of the time. The walls were built of neatly trimmed stones, but a

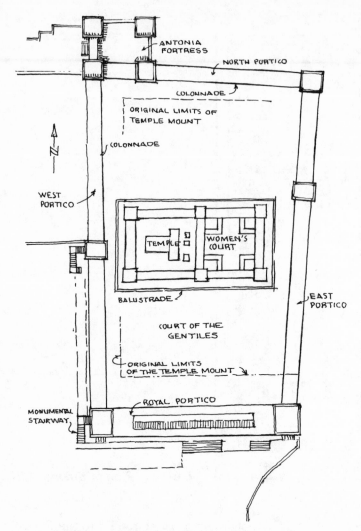

Figure 32: Plan of the Second Temple Mount in Jerusalem
In 18 B.C. Herod the Great extended the original limits of the Temple Mount
to the north and south of where they had been at the time of Solomon's
Temple. Major structures on the mount included the esplanade; the
porticos along the perimeter; the Antonia Fortress, at the northwest
corner; and the Temple area (Canaan 1975, 58, 59).

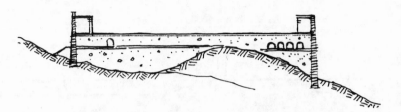

Figure 33: Temple Mount Section Through the South End
When he rebuilt the Temple, Herod the Great constructed retaining walls
of enormous ashlars up to thirty-six feet in length. They rest on the
terrain's natural bedrock, which is fifty to sixty feet deep and provides
excellent support for such a heavy structure. Between the bedrock and
esplanade, builders supported the construction with backfill or created
stone-arch storage rooms or reservoirs.

Inside these walls may lie the remains of those built by the Jebusites,
King David, King Solomon, and the Maccabees.

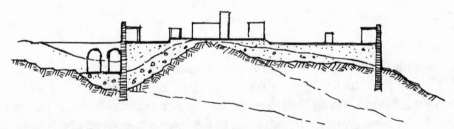

Figure 34: Temple Mount Section Through Mount Moriah
To the left of the outcrop of Mount Moriah lies the filled-in Tyropoean
Valley, to the west of the mount. Also shown are the Temple, the west and
east porticos, and the Robinson Arch.

Although the Bible fully describes the building and ornaments of
Solomon's Temple and palace, it provides us with little information
concerning the retaining walls and enclosure.

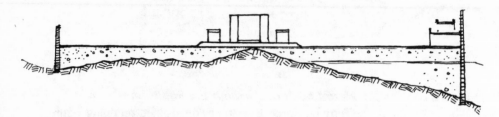

Figure 35: Longitudinal Section Through the Temple
The rock outcrop that projects through the esplanade or platform is
believed to have been the location where Abraham was ready to offer his
son to God. Generally this is also regarded as the precise location of
Solomon's Temple. Today the Dome of the Rock stands in its place.

On the right, along the south wall, stands the Royal Portico. Bedrock,
approximately 150 feet below the esplanade, forms the wall's foundation.
The deck of the platform lies on man-made fill.

beam of cedar was placed at every third course, strengthening the walls and
providing support for the interior paneling (Hamlin 1904, 4).

The limestone blocks had been quarried nearby, hauled on rollers, and
hoisted by wooden cranes. Though most of the blocks measured about three feet
wide and sixteen feet long, some of the blocks for the retaining walls were up to
thirty-five feet in length. One of Jesus' disciples remarked on the awesome beauty
of the Temple: ". . . Look, Teacher! What massive stones! What magnificent
buildings!" (Mark 13:1 NIV).

The Temple Mount stood within the ancient city of Jerusalem, which was
about 3.5 miles in circumference and had about 25,000 inhabitants. By the west
retaining wall of the Temple Mount lay a ravine, crossed by masonry bridges and
stairways. (Parts of these structures are now known as the Wilson Arch, the
Warren Arch, and the Robinson Arch.) Further west was the Lower City. South of
the Temple Mount were the old City of David (Ophel), the Kidron Valley, and
parts of the Lower City. To the north lay the Upper City, with its theaters and
wealthy homes. To the east loomed the Mount of Olives (La Perrousaz 1987,
36, 37).

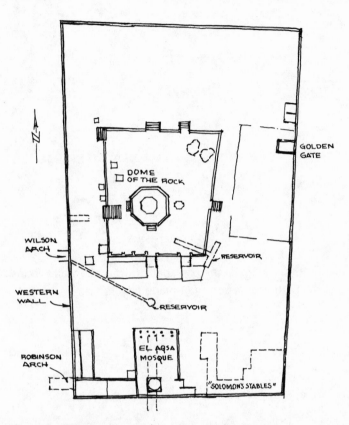

Figure 36: Present-Day Plan of the Temple Mount
Today the Dome of the Rock (Mosque of Omar), which stands directly over
Mount Moriah, and the El Aqsa Mosque, on the south wall, dominate the
Temple Mount. Arched rooms under the southeast esplanade form
Solomon's Stables. A major reservoir and aqueduct lie near the Dome of
the Rock.

  Without plans for the Temple Mount that was built during Solomon's
reign, archaeologists cannot know for certain where his workmen built the
original retaining walls, but the present Temple Mount certainly covers
more area than it did at the time of Solomon.

  Various theories concerning the walls have been proposed. One holds
that the present east wall dates back to the time of Solomon, because a
continuous vertical joint exists about 107 feet north of the mount's
southeast corner (Cornfeld 1981, 317).

6
Ancient Building Materials

Go back two or three thousand years and you find builders using essentially the same materials as we use today. Of course, without the aid of electric- or gasoline-powered equipment, they had to manufacture and assemble the materials by hand. But just like modern contractors, they worked with wood and stone and cement. The only significant modern structural material not available to them was prefabricated steel beams—yet they regularly used iron fittings to connect timber and masonry.

Geologists tell us the Holy Land was once at the bottom of a sea. The skeletons and shells of microscopic sea animals sank to the sea floor and were compressed, over the millennia, by the collecting sediment, forming white limestone. When the earth's own turbulence forced this sea floor up above sea level, it formed mountains of limestone and dolomite. As wind, water, and volcanic action eroded this material, it formed various kinds of soils and exposed rock.

These processes, in a way, stocked a "warehouse" full of materials for ancient builders. We have seen how Solomon had limestone quarried for the Temple. But the ancients also used the soils—sand and clay. Other minerals and metals were mined and developed for use in construction—gypsum, pozzolana, iron, and so on. Of course, trees were also cut and milled into usable lumber.

Builders would dig sand from sand pits, washing it to remove dirt and salt, and then use it in mortar and plaster (beach sand was much too rounded to cohere well). With charcoal fires, they would heat the ores of lime, gypsum, pozzolana, and metals, to convert them to construction material.

Clay makes great bricks: just add water. Throughout the Mediterranean world, workers would dig up clay mud, add straw fibers, to increase tensile strength, trample it to the proper consistency, shape it with wooden molds; and dry these bricks in the sun. Baked or fired brick, though long known, did not come into general use until the late Roman Republic. According to Vitruvius, mud, sun-dried brick had been declared illegal for house walls in Rome, creating a need for kiln-fired clay bricks. Roman bricks came in many shapes and sizes, but the Romans seemed to prefer long, wide bricks that were only an inch and a half thick. These were less likely to warp or crack than thicker bricks.

Stone Masonry

Throughout the ancient world, builders used the stones available to them. Construction materials included limestone, marble, sandstone, basalt, granite, and gneiss, among others.

Limestone was a nearly perfect material for construction in central Palestine. Plentiful in that area, it was soft and easy to quarry, but it hardened once it became exposed to the environment. To quarry the stone, masons drilled several holes into the rock in a straight line and drove wooden pegs into the holes. When soaked with water, these pegs expanded, exerting pressure on the stone and splitting it in a rather straight line. Using chisels and adzes, masons then shaped the rough faces of the stone. With square plumbs, levels, and measuring strings, they ensured the straightness and angles of the stone block, which would then be sanded with a loaflike rubbing stone. Other stones were quarried in similar ways.

The stones used in any given locale depended on the geology of that area and the ability to quarry it. We have seen how the early Egyptians used sandstone for the bulk of the pyramids, because that was close at hand. The Babylonians, in the fertile plains of Mesopotamia, were more dependent on clay brick.

The Romans first used tufa, a soft tan and brown volcanic rock, in stone-masonry construction. Later they learned to use the harder *lapis gabinus* and *lapis albanus* (now called sperona and peperino), formed by the action of water on a mixture of volcanic ash, gravel, and sand. Still later, they used a hard, attractive limestone known as travertine (*lapis tiburtinus*). But unlike the volcanic rocks, it was not fire resistant; under heat, it crumbled into powder (Morgan 1960, 49).

Figure 37: Brickmaking
This simple method of making bricks was common around 1500 B.C., in
Egypt, Crete, and Asia Minor. Clay was mixed with water and straw and the
mixture was carried and poured into wooden molds, which were then lifted
to allow bricks to dry in the sun.

Gypsum

As early as 3000 B.C., the Egyptians lined the inside of their pyramids with gypsum
plaster. The Greeks also used a gypsum coating in their temples, and the Romans
used it in their houses.

In Greek the word *gypsos* means "plaster." A light-density gray-white rock,
it can be ground to a powder, from which plaster of Paris is made. Chemically,
the rock is known as dihydrous calcium sulfate ($CaSO_4 \cdot 2H_2O$). Heated in pits,
gypsum gives up three-quarters of H_2O as water vapor. As a result of this "calcining," it can be easily crushed into a talclike powder. When water is added
again and it is mixed, it returns to its original rocklike hardness.

Mortar

The Israelites made mortar by mixing lime, sand, ashes, and water into a plaster.
This was watertight enough to coat cisterns and reservoirs or for use as a finish
coating for clay masonry walls and roof decks.

The Romans also used lime mortar between bricks and stone blocks on
occasion. But in their public buildings, they generally relied upon iron clamps

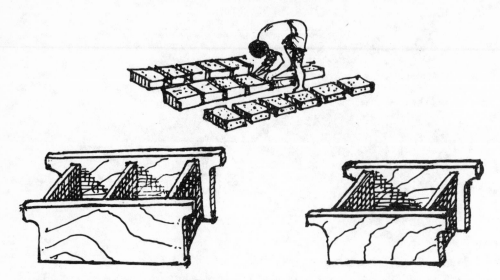

Figures 38 & 39: Brick Molds
The man above is making bricks with a mold. Some molds made one brick, others made two (*below*). They often had handles to facilitate lifting (Miller 1978, 33, 34).

and the close fit of the neatly trimmed stones to hold the structures together (Morgan 1960, 45).

Concrete

Near Vesuvius and elsewhere in Italy were deposits of a sandy volcanic ash. When added to lime mortar, this made a cement that would harden even under water. They named this material *pulvis puteoli,* after Puteoli, an area that contained huge beds of it. By mixing this cement with sand and gravel, the Romans created a sturdy concrete.

But initially the Romans apparently did not realize the possibilities of this new concrete, because they used it only in small amounts. Builders continued to apply mortars and plasters for centuries, until the new waterproof concrete began to catch on.

Vitruvius described in detail the importance of the quality of sand used in mortar and concrete:

> In walls of masonry the first question must be with regard to the sand,
> in order that it may be fit to mix into mortar and have no dirt in it. The kinds

Figures 40 & 41: Building a Masonry Wall
Dressed stones would be carefully placed, sometimes cemented with
mortar, and masons would plumb the wall with simple, but effective,
equipment.

Figure 42: A Stone Rubble Masonry Wall
In a rubble stone wall, undressed, randomly shaped, uncoursed stones
were set in naturally stable positions. This was also known as polygonal or
cyclopean masonry.

of pit sand are these: black, grey, red, and carbuncular.... The best ...
crackles when rubbed in the hand, while that which has much dirt in it will
not be sharp enough....

But if there are no sand pits from which it can be dug, then we must sift
it out of river beds or from gravel or even from the sea beach. This kind,
however, has these defects when used in masonry: it dries slowly ... and such
a wall cannot carry vaulting....

But pit sand used in masonry dries quickly, the stucco coating is permanent, and the walls will support vaulting. I am speaking of sand fresh from
the sand pits. For if it lies unused too long after being taken out, it is disin-
tegrated by exposure ... and becomes earthy.... Fresh pit sand, however, in
spite of all its excellence in concrete structures, is not equally useful in
stucco, the richness of which, when the lime and straw are mixed with such
sand, will cause it to crack as it dries on account of the great strength of the
mixture. But river sand, though useless in "signinum" on account of its thin-
ness, becomes perfectly solid in stucco when thoroughly worked by means of
polishing instruments (Morgan 1960, 44).

Timber
Wood has been used in building since prehistoric times. Tree limbs and branches
served as beams for the roofs and walls of many early homes.

Figure 43: Quarrying Stone
Stone masons cut and dressed blocks of stone in the quarry. The mason in the center is drilling notches into the rock. On the left, another man drives wooden pegs into such notches. When he soaks the pegs with water, they will expand, splitting the rock. The mason at the right finishes a cut block's surface with a rubbing stone (Miller 1978, 359–361).

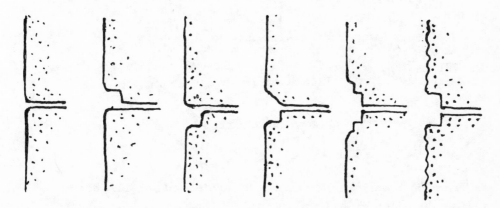

Figure 44: Types of Finished Joints in Stone Masonry

The Bible indicates that parts of Palestine were rich in timber (1 Kings 10:27). The common people used sycamore, an inexpensive and readily available wood. The wealthy often imported cedar, fir, or almug from Lebanon or Syria. In public buildings, made mostly of stone, timber was generally used for the roof and doors.

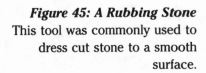

Figure 45: A Rubbing Stone
This tool was commonly used to
dress cut stone to a smooth
surface.

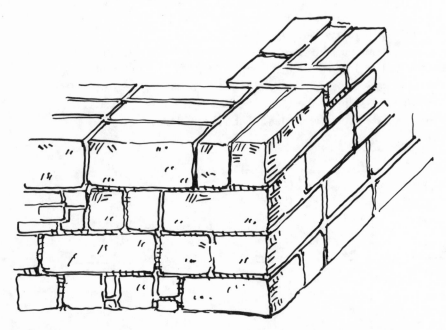

Figure 46: A Stone Wall Consisting of Header and Stretcher Blocks
(Paul 1973, 30, 31).

King Solomon contracted with King Hiram of Tyre to import "cedars of
Lebanon" for the Temple construction (1 Kings 5:6). Thirty thousand Israelite
lumberjacks were drafted and sent to Lebanon—ten thousand at a time, in
one-month shifts—to work with the local woodsmen in felling trees for the
Temple. The cedars and pines were then floated south in rafts on the Mediter-

Figure 47: Plastering a Masonry Wall
A mason coats an ashlar or cut-stone masonry wall with lime plaster, using
a wood pallet for mixing, a water jug for thinning, and a trowel for
application and finishing.

ranean Sea. Israelite "carriers" apparently transported the timber inland to Jerusalem.

In his *Ten Books of Architecture,* Vitruvius offered numerous recommenda-
tions for selecting and preparing timber:

> Timbers should be felled between early Autumn and the time when
> Favonius begins to blow. For in spring all trees become pregnant, and they are
> employing their natural vigor in the production of leaves and of the fruits that
> return every year.
> In felling a tree we should cut into the trunk of it to the very heart, then
> leave it standing so that the sap may drain out drop by drop throughout the
> whole of it. In this way, useless liquid which is within will run through the
> sapwood instead of having to die in the mass of decay, thus spoiling the
> quality of the timber. Then and not till then, the tree being drained dry and the
> sap no longer dripping, let it be felled and it will be in the highest state of
> usefulness (Morgan 1960, 58).

Metals

You could almost date the birth of civilization to the discovery of metallurgy. Ancient cultures graduated at different times, in different ways, from the Stone Age to the Bronze Age. The early chapters of the Bible even mention Tubal-Cain, "who forged all kinds of tools out of bronze and iron" (Genesis 4:22 NIV). Ever since, people have used metals for tools and weapons.

As early as the third millennium B.C., the Egyptians opened up copper mines in the Sinai Peninsula. The authorities sent their prisoners of war to work in these mines, digging with bare hands, hammers and wedges, bones, and horns. They used fire and vinegar for blasting.

Tin, which along with copper makes up the bronze alloy, was not plentiful in the Middle East, though sometimes it was smelted from richer ores (Ezekiel 22:18). Phoenician merchant ships brought tin and other metals to the Middle East from Tarshish—probably Spain (Ezekiel 27:12).

The earliest known use of bronze in the biblical region occurred in Sumer, in Mesopotamia, about 3000 B.C. The Sumerians had the ability to smelt copper and alloy it with small quantities of tin and arsenic. This process formed bronze, which was harder than copper yet could be cast in molds. Bronze became the dominant metal in the Middle East until iron superseded it, about 1200 B.C.

The Bible notes extensive use of bronze for Temple items: ". . . All these objects that Huram made for King Solomon for the temple of the Lord were of burnished bronze. The king had them cast in clay molds in the plain of the Jordan between Succoth and Zarethan. Solomon left all these things unweighed because there were so many . . ." (1 Kings 7:45–47 NIV).

Early Metal Molding

When ancient metalworkers realized that copper (and later bronze) melted at high temperatures, they learned to shape the metal by pouring it into molds of clay or stone. The first molds were simple depressions cut into stone or molded in clay, and the metal objects cast in such molds required much hammering and polishing.

Later metallurgists learned to use two opposing molds, avoiding much of the hammering and finishing work. Two-piece molds were crafted from stone or fired clay and held together by short dowels.

Through the second and third millennia B.C., bronze was widely used to make cups, urns, and vases; farming tools as well as primitive stoves; and weaponry, such as knives, spears, helmets, and shields.

Fuels

In addition to the more common uses of fire in the ancient world—cooking and heating—smelting furnaces and the kilns that baked bricks required fuel.

Wood and charcoal were the most common fuels. For cooking, most preferred the slow-burning charcoal. Being almost pure carbon, it could reach extremely high temperatures, also making it ideal for smelting furnaces.

The best charcoal came from the hardest and closest-grained wood, such as oak and beech. Trees would be cut into small sections and partially burned in pits. These pieces would then be chopped into smaller pieces and bagged for shipment.

Iron

The Roman historian Pliny described iron as "the most useful and the most fatal instrument in the hand of man." Harder than bronze, it rapidly made other tools and weapons obsolete. The passage from bronze technology to iron technology had significant geopolitical effects throughout the ancient world.

The Iron Age dates to about 1200 B.C., though it hit different areas at different times. Ironworking probably arose first in the Aegean. The seafaring Phoenicians may have brought these skills to Canaan, where iron rapidly became the monopoly of the Philistines.

Canaan had plenty of iron. In Deuteronomy 8:9 (NIV), Moses referred to the Promised Land as ". . . a land where the rocks are iron and you can dig copper out of the hills." But only the Philistines knew how to work with iron, and they safely guarded their technology for a couple of centuries. When the Israelites were having trouble in their conquest of the land, some complained to Joshua, ". . . All the Canaanites who live in the plain have iron chariots . . ." (Joshua 17:16 JFB). First Samuel 13:19–22 (TLB) illustrates the Philistines' continuing dominance in metallurgy in the eleventh century B.C. and the effect on the neighboring Israelites:

> There were no blacksmiths at all in the land of Israel in those days, for the Philistines wouldn't allow them for fear of their making swords and spears for the Hebrews. So whenever the Israelites needed to sharpen their plowshares, discs, axes, or sickles, they had to take them to a Philistine blacksmith. (The schedule of charges was as follows:

ANCIENT KILNS AND FURNACES

Figure 48: Ancient Kiln

Figure 49: A Potter's Oven at Tell Qasileh

In 3500 B.C. Mesopotamian kilns were constructed with a perforated floor over the firebox. In the chamber at right, the potter placed items for baking. This prevented ash and pieces of wood from coming in contact with the surfaces of the vessels, causing defects.

Figure 50: Section Through a Copper Casting Furnace

A beehive-shaped furnace with a stone stack was placed over a trench, to admit combustion air. Clay crucibles containing molten copper stand over a charcoal fire. Air holes in the bottom of the furnace create a draft or provide openings for use of bellows (Cornfeld 1976, 161).

Figure 51: A Beehive-Shaped Furnace

Figure 52: An Ancient Egyptian Furnace (Cornfeld 1976, 16).

For sharpening a plow point, 60¢
For sharpening a disc, 60¢
For sharpening an axe, 30¢
For sharpening a sickle, 30¢
For sharpening an ox goad, 30¢)

So there was not a single sword or spear in the entire "army" of Israel that day, except for Saul's and Jonathan's.

It was only a matter of time before the Israelites learned how to work iron. In David's reign we read of iron axes (2 Samuel 12:31) and of iron nails and fittings stockpiled for the construction of the Temple (1 Chronicles 22:3).

Iron Making. Making iron was relatively simple. Iron ore was crushed and placed in a deep bed of burning charcoal. Carbon from the charcoal would combine with oxygen from the iron ore (iron oxide) and escape as carbon dioxide gas, leaving the iron relatively pure.

Although early metallurgists used hollow tubes to blow on their fires and make them hotter (figure 52), and later ones used bellows, they could not heat the iron to its melting point (about 1,530 degrees centigrade), but they could soften it into a workable, spongy mass called a "bloom." This would still have silica impurities trapped inside, requiring repeated heating and pounding to purify it. Eventually the pure metal would be softer than bronze and could be pounded into a usable shape.

This was "black iron," fairly strong, but still limited in several ways. Forging was a difficult way to purify the metal—and it still retained many impurities. Since it required the hands-on work of pounding and forging, only a small amount of iron could be worked with at a time. The Bible depicts the wearisome work of the blacksmith:

The blacksmith takes a tool and works with it in the coals; he shapes an idol with hammers, he forges it with the might of his arm. He gets hungry and loses his strength; he drinks no water and grows faint.

Isaiah 44:12 NIV

So it is with the smith, sitting by his anvil, intent on his iron-work. The smoke of the fire shrivels his flesh, as he wrestles in the heat of the furnace. The hammer rings again and again in his ears, and his eyes are on the pattern

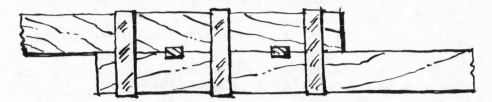

Figure 53: Timbers Spliced With Oak Keys and Yoke Straps

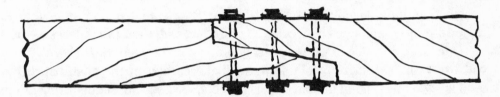

Figure 54: A Lap-and-Scarf Butt Joint
This is keyed together with hardwood and anchored with fishplates and bolts

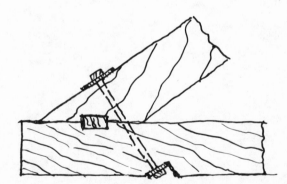

Figure 55: A Butt Joint Anchored With a Key Bolt and Washer

he is copying. He concentrates on completing the task, and stays up late to give it a perfect finish.

Ecclesiasticus 38:28 NEB (Apocrypha)

When ironworkers gained the ability to make hotter fires, it greatly improved their product. Black iron was smelted at 800–1,000 degrees centigrade. But at 1,530 degrees centigrade, the iron liquefied. Much freer from impurities, this iron could also be cast in molds.

Not only did iron revolutionize warfare, with iron-plated chariots and weapons, and farming, with iron spades and plows; it also made masonry and timber construction more feasible. Iron shafts, nails, dowels, plates, and other fittings created strong connections between other building materials. (*See* figures 53–55.)

One more advance in ironworking should be mentioned. Heated with charcoal, iron can "carburize," combining with the carbon in the charcoal. If the metal is suddenly dipped in cold water, it becomes steel. Subsequent reheating, or tempering, reduces this metal's brittleness. Some modern scholars theorize that the Spartans owed their success in their fifth-century B.C. struggles against Persia and Athens to their invention of steel. Their steel weapons would easily outclass the softer wrought-iron or bronze weapons of their foes (Muhly 1982, 42–50).

7
Trades, Tools, and Machines

The quality of the ancient infrastructures—the buildings, roads, waterworks, vehicles, and so on of society—depended on three factors. First, the builders needed the proper skills. We have already seen how iron was plentiful in Canaan, but the Israelites did not know how to forge it into useful implements: They lacked crucial skills.

Second, the workers needed proper tools. Stone masons required hammers, chisels, plumb lines, and levels. Carpentry developed with the advent of adzes, awls, saws, and metal nails. Metallurgists needed furnaces that could reach high temperatures, often with the aid of bellows, and strong hammers and anvils to forge their products. Tools needed to be strong enough and accurate enough for the tasks at hand.

Finally, the history of civilization shows us the importance of new machinery. Many of the great advances in building and weaponry and transportation came about through the invention of new contraptions that got the job done better. Starting with such simple devices as the block and tackle, derricks, gin poles, levers, and screw jacks, certain societies got the jump on others through clever use of new machines. The pattern has repeated throughout history.

Quarrying and Stone Masonry

Generally, in biblical times, the craft of masonry, dressing and placing stones, also involved quarrying. In fact, the stones were usually hewn into shape at the quarry.

Stone masonry had its heyday in Israel during the building of Solomon's Temple.

ANCIENT TOOLS

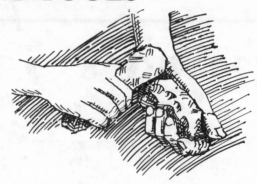

Figure 56: Shaping a Flint Tool
Flint, the use of which dates back to the Lower Paleolithic period, was probably the first tool used by humans. Flakes would be chipped off a suitable stone, leaving a rough, dimpled surface that could be used for cutting, chopping, or scraping.

Figure 57: An Iron Tool
(Reader's Digest Editors *People* 1981, 116).

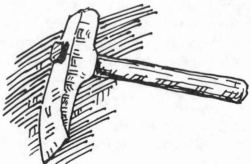

Figure 58: An Early Hammer

Figure 59: An Early Adz

Figure 60: A Plowshare
This tool was found in Megiddo during the
excavations by Gottlieb Schmacher in 1903–
1905. The blacksmith's workshop in which it
was found dates from about the tenth century
B.C. (Muhly 1982, 54; Reader's Digest Editors
People 1981, 116).

Figure 61: A Hoe
This tool was also found at Megiddo from
the same period.

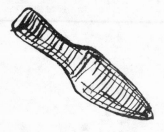

Figure 62: An Ox Goad
A third tool was found in the blacksmith's workshop
in Megiddo (Reader's Digest Editors *People* 1981,
116).

Figure 63: An Iron Tool
      (Muhly 1982, 45).

Figure 64: An Iron Pick
This tool was found in Har Adir, in the Upper Galilee, and is from the
twelfth or thirteenth century B.C. Laboratory testing of the metal indicated
that the tool was made of steel—carburized iron that had been quenched
and then tempered.

Figure 65: A Biblical Plow
This bronze plowshare and wood tailpiece are similar to those farmers in the Holy Land use today.

Figure 66: A Level

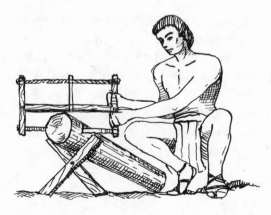

Figure 67: A Carpenter Using an Ancient Saw
A wood strut separated the saw's two arms, which held a metal blade at one end and a twisted rope tie at the other. The tie was tightened by twisting the rope with a tourniquet (Reader's Digest Editors *People* 1981, 329).

> Solomon had seventy thousand carriers and eighty thousand stonecutters in the hills, as well as thirty-three hundred foremen who supervised the project and directed the workmen. At the king's command they removed from the quarry large blocks of quality stone to provide a foundation of dressed stone for the temple.
>
> 1 Kings 5:15–17 NIV

Solomon commissioned other building projects besides the Temple. His own palace took thirteen years to complete and was "... made of blocks of high-grade stone cut to size and trimmed with a saw on their inner and outer faces" (1 Kings 7:9 NIV).

Of course, the Egyptian pyramids, built much earlier, remain the crowning feat of stone masonry. Those huge stones also tell us more about the craft. Marks on the stone indicate the use of straight saws, circular saws, and tubular drills with jeweled cutting points. The bronze saw blades were .003 to .002 of an inch thick.

But stone walls also cross the Judean hills, creating terraces for farming. These walls, which trap the rainfall and retain the soil, represent the work of some of the earliest stonemasons.

Carpentry

The New Testament identifies Jesus as a carpenter (Mark 6:3) and the son of a carpenter (Matthew 13:55). Elsewhere, the Bible frequently mentions carpenters and their tools—dating back to Noah (Genesis 6) and Bezalel, who made the Ark of the Covenant (Exodus 37). Isaiah describes the carpenter's craft: "The carpenter measures with a line and makes an outline with a marker; he roughs it out with chisels and marks it with compasses . . ." (Isaiah 44:13 NIV).

Later, Jeremiah ironically refers to the hammer and nails a pagan carpenter would use to keep a wooden idol from falling down (Jeremiah 10:4).

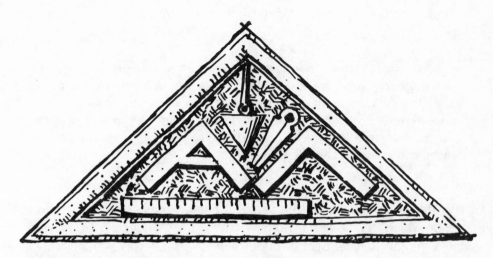

Figure 68: A Carving Showing Roman Tools
This stone carving, found in a tomb of the first century A.D., shows common
tools used by the Roman craftsmen, including a ruler; an adjustable
triangle; a plumb bob, or plummet; a divider, compass, or caliper; and a
square set.

Machines

Beginning with simple devices—pulleys and levers—builders of the biblical period developed machines that would enable them to do more, bigger, and better
things.

Wheels and Axles. The earliest known use of wheel was the potter's wheel,
invented in Mesopotamia about 3500 B.C. This flat disc of wood or stone had the
center hollowed out on the underside, so it could fit over a stone or hardwood
dome set in the ground. The wheel and axle soon followed, revolutionizing
transportation. Subsequent developments included the waterwheel and the
geared windlass (Landels 1978, 20).

Pulleys. By using a pulley, an engineer can reduce the amount of force required
to move a heavy load. A two-block pulley, known as a block and tackle, provides

even more advantage. The more pulley wheels applied to a job, the greater the mechanical advantage.

A third-century A.D. account describes the pulley's use:

> If we want to move any weight whatever, we tie a rope to this weight and
> ... pull the rope until we lift it. And for this is needed a power equal to the
> weight that we want to lift. But if we untie the rope from the weight, and tie
> one of the ends to a solid cross beam and pass its other end [around] a pulley
> fastened to the middle of the burden and draw on the rope, our moving of that
> weight will be easier (Landels 1978, 4).

Levers. Archimedes extolled the value of simple lever, crowing, "Give me a fulcrum on which to rest, and I will move the world." This early Greek engineer explained the mechanical operation of the lever in this way: "Equal weights at unequal distances . . . incline toward the weight which is at the greater distance." Thus, ancient builders learned they could move large objects easily, with a well-positioned lever (De Camp 1963, 156).

Screws. About 200 B.C. a Greek mathematician, Apollonius of Perga, worked out the geometry of the spiral helix, on which the screw is based. This concept, employed in a screw jack, can be used to lift heavy objects.

Lifting Devices

Rope. Since ancient times, people have used vegetable material for hauling and lifting loads. First, vines were twisted together to form stronger strands that were still flexible enough for tying knots. Other rope materials included the fibrous bark of the palm tree, camel hair, horsehair, leather thongs, jute, sisal, flax, and wild and cultivated hemp.

PULLEYS

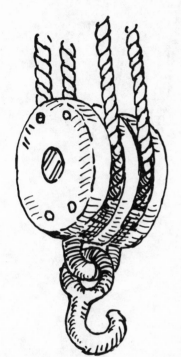

Figure 69: A Parallel Two-Sheave Block

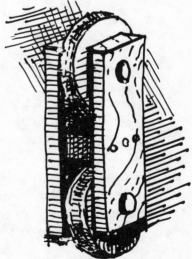

Figure 70: A Tandem Two-Sheave Block
The side plates on the sheaves are called the
cheeks.

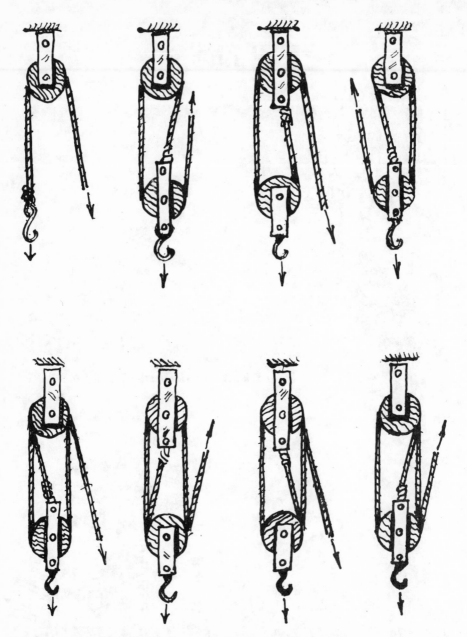

Figure 71: Various Block-and-Tackle Arrangements
The mechanical advantage of each of these pulley arrangements is directly
proportional to the number of pulleys involved. Arrows represent the applied pull
(Landels 1978, 84).

Figure 72: Gin Pole
This ancient form of rigging employed the pulley concept to haul building
stones. Its parts included the mast, which was held in position by guy lines
and anchor pegs (or "dead men"); a three-part load line; a two-sheave
pulley block; and a single-sheave block (De Camp 1963, 26).

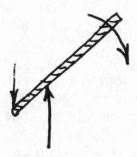

Figure 73: Forces Involved With Leverage
A relatively small force at the end of a lever can result in
a large force near the pivot point, or fulcrum (Rossnagel
1964, 10).

Figure 74: Ancient Flaxen Rope Making
The manufacture of flaxen rope required a team of three in a "rope walk."
As one man walked forward, twirling weighted strands, another, at the
center, controlled the tautness of the rope with spikes; the third man
combined the strands by twirling in the opposite direction (Reader's Digest
Editors *People* 1981, 233).

After cutting individual fibers into roughly seven-foot lengths, rope makers
drew them out on frames, which combed out the fibers and made them parallel.
Cordage, or special oils, applied in this stage, lubricated and preserved the rope.

The fiber, or slivers, came out in a fluffy stream, and were recombed six or
seven times. During this process, rope makers blended in other fibers of the
same grade to ensure a uniform quality. They then spun the fibers into a yarn.
This would be twisted, left-handed, into a strand. Three or more strands would
then be twisted, right-handed, into the finished rope.

Ancient mariners preferred flaxen rope, as did many builders, hunters, and
warriors. Its manufacture involved a team of three in a "rope walk." The rope
strands would be stretched tight, with a man at each end and a third in the
center. The end of each strand was attached to a tool that was weighted by a
stone. At one end a man would begin twirling these stone-laden tools, twisting
the strands, while walking toward the center. The man at the other end would

Figure 75: A Simple A-frame

Figure 76: An A-frame and Windlass
This simple hoisting rig included a timber A-frame, fastened by an iron bolt at the apex and held upright by ropes. From this was suspended a double-pulley block and a lower single-pulley block. The hoisting rope was wound by a windlass (De Camp 1963, 26).

back away, twirling in the other direction, while the center man tightened the strands with a spike. Thus the individual fibers twisted in one direction, forming two strands, and these strands were twisted in the opposite direction, forming a single rope (Rossnagel 1964, 14, 15).

Cranes and Capstans. As long as the transportation of building materials relied on simple manpower or animal power, the size of the materials was limited. Apart from monumental tasks like the pyramids, most building stones needed to weigh less than eighty pounds or so. For instance, the columns in early Greek temples (fifth century B.C.) consisted of column drums piled one on top of the other. The weight of monolithic columns would have been prohibitive.

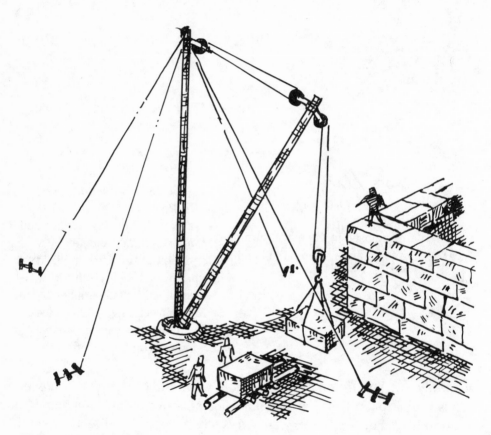

Figure 77: A Derrick

But then two popular mechanical devices—the capstan (or windlass) and the crane—changed all that. Ropes and various combinations of pulleys could now multiply the effective force of the workers.

The windlass used the winding of a rope around a rotating drum to exert force for lifting. At first, handspikes inserted into the drum did the winding. Later workers used cranks, which generally had less power but greater speed. Sometimes the handspikes were long enough that several men could turn simultaneously, developing great power. To prevent a backlash, the windlass sometimes had a ratchet-type device that kept the cylinder from rotating backwards.

Vitruvius described the set-up and workings of the basic A-frame and windlass-pulley system:

> Two timbers must be provided, strong enough to withstand the weight of the load, and they must be fastened together at the upper end by a bolt; then they are spread apart at the bottom, thus set up and kept upright by ropes attached to the upper ends and fixed at suitable intervals all around.
>
> A block is fastened at the top [called a "rechamus"]. Two sheaves [pulleys] are enclosed in the block, turning on axles. The traction rope is carried over the top sheave, then let fall and passed around the sheave in the block below. It is then brought back to the sheave at the bottom of the upper block and thus goes down to the lower block; it is fastened there through a hole in the block.
>
> The other end of the block is pulled back and down through the legs of the hoisting machine.
>
> Socket pieces are nailed to the hinder faces of the squared timbers at point where they are placed apart; the ends of the windlass are also inserted into them. Two holes are made close to the windlass so that the axles may turn freely and so adjusted that the handspikes can be fitted into them.
>
> Shears made of iron are fastened to the bottom of the lower block, and their prongs are brought to bear upon the stones. The latter must have suitable holes bored into them. As one end of the rope fastens to the windlass—and the latter is turned around by working the hand spikes—the rope winds around the windlass, becomes tightly drawn, and thus raises the heavy loads to the heights desired in the work.
>
> . . . When we must provide machines for heavier loads, we should use timbers of greater length and thickness, and be sure to provide them with corresponding large bolts at the top, as well as windlasses for turning at the

Figures 78 & 79: Ancient Treadmills

To provide power for screw pumps, ancient man used treadmills. Above you see the treadmill alone. The complete device included a platform, or base, an inclined jib with a bracket that held the rotating shaft, a drum, and the treadmill (figure 79, *below*) (De Camp 1963, 193).

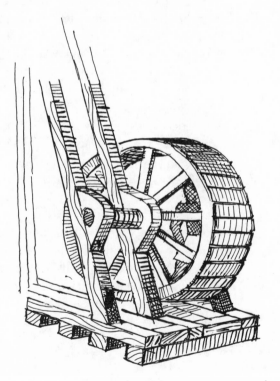

Figure 80: A Millstone for Crushing Olives
With this device a man or animal could crush or grind olives or other materials to extract oil. As it rolls around the basin, the stone crushes and squeezes the fruit.

bottom. When these are provided, let forestays be attached, then left lying slack in front; carry the backstays over the shoulders of the machine—to some distance; if there is nothing to which they may be fastened, sloping piles may be driven, the ground rammed down hard all around to fix them firmly, and the ropes made fast to them.

A block should then be attached by very stout cord to the top of the machine and from that point a rope be carried to the pile. The rope should be put around the sheave of this block, then brought to the block that is fastened to the top (of the machine). The rope should be passed around its sheave, and then go down from the top—and back to the windlass, which is at the bottom of the machine; it should then be fastened.

The windlass must now be turned by means of hand spikes, and it will raise the machine without any danger. A machine of the larger type will now be set in position, with its ropes in their place about it, and its stays attached to the piles (Morgan 1960, 295).

Treadmills. The treadmill, or treadwheel, further harnessed human and animal power for lifting, turning, and pulling. By climbing the "rungs" of a vertically

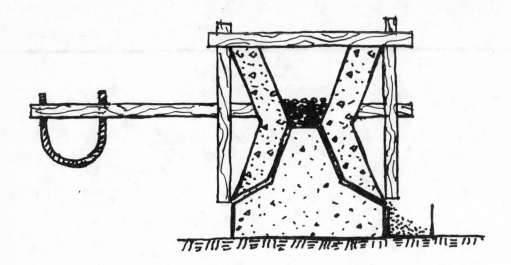

Figure 81: A Donkey Mill

Usually a donkey would power this mill, thus the name it bears. Into the
upper cone the raw material was loaded. As the donkey turned the mill,
the material would fall between the faces on the lower parts of the cone.

The modern-day gyratory crusher, used in mills to reduce the size of
ore, is a more sophisticated version of this mill (De Camp 1963, 244).

placed wheel, a man could use his own weight to turn the wheel and pull or hoist
whatever it was attached to. The treadmill also offered greater mobility. A
combination treadmill and crane could be dismantled—the jib could be laid
horizontally on one or more carts, and the wheel could be rolled along the
roadway.

With such machines, ancient builders hoisted huge blocks—ashlars weighing
tens of tons—and precisely placed them in position. While some earlier wonders
of construction, such as the pyramids and Solomon's Temple, used thousands of
forced laborers, even these builders used primitive machines and methods to multiply the effective force of their manpower. As time went on, machines enabled
fewer workers to do more work. As early as the first century A.D., the Roman Emperor Vespasian feared that mechanization might cause unemployment:

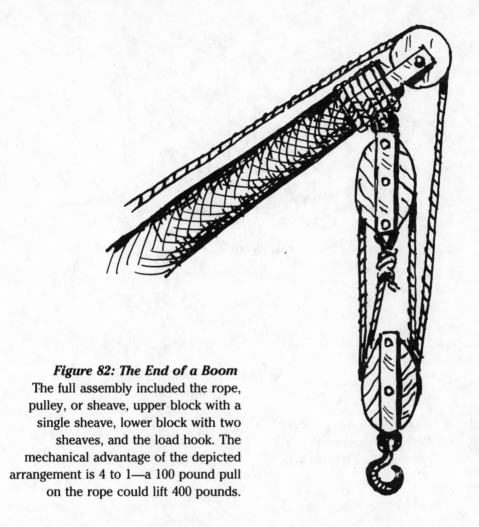

Figure 82: The End of a Boom
The full assembly included the rope,
pulley, or sheave, upper block with a
single sheave, lower block with two
sheaves, and the load hook. The
mechanical advantage of the depicted
arrangement is 4 to 1—a 100 pound pull
on the rope could lift 400 pounds.

An engineer offered to haul some huge columns up to the Capitol at a
moderate expense by a simple mechanical device, but Vespasian declined his
services, saying, "I must always ensure that the working class earn enough
money to buy themselves food." Nevertheless he paid the engineer a very
handsome fee (De Camp 1963, 170).

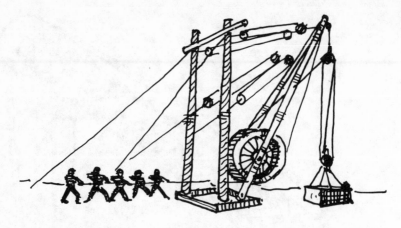

Figure 83: A Guyed Double-Mast Derrick

Figure 84: A Guyed Single-Mast Derrick, Powered With a Treadwheel

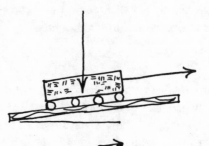

Figure 85: Forces Involved in Transporting Heavy Stone Blocks

The force required to move the load up a slight incline is only a small fraction of the weight of the stone.

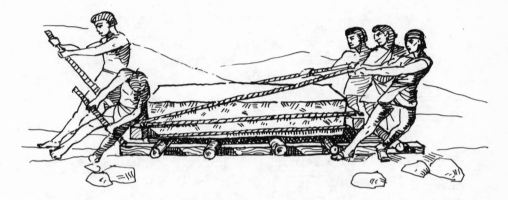

Figure 86: One Method of Moving Large Stone Blocks

With moveable wooden rails and rollers, set progressively under a stone block, workmen may have moved each block of limestone that made up Cheops's pyramid; blocks weighed up to 2.5 tons. As the men pulled the ropes, they applied leverage to one end of the stone (Reader's Digest Editors *People* 1981, 359).

Figure 87: One Theory About Erecting an Obelisk.

Figure 88: How Obelisks May Have Been Transported

The workers pull on ropes attached to the stone shaft. Below the stone lie wooden rails and rollers, dampened with water to reduce friction. As foremen clap their hands or beat on drums, the workers rhythmically pull their lines.

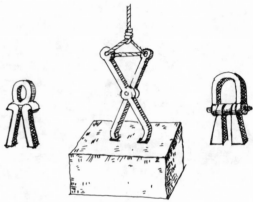

Figure 89 & 90: Ancient Hook Devices

About 800 B.C. the Greeks used these hooks for lifting heavy stone blocks. Mortises, or bored holes, cut into the stone face, allowed the use of various types of anchors. The complete assembly included the lower load block, iron shears or forceps, and prongs (Landels 1981, 84).

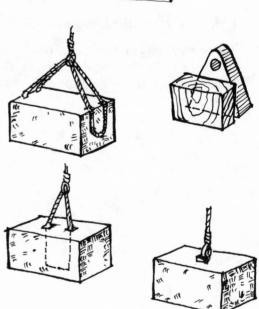

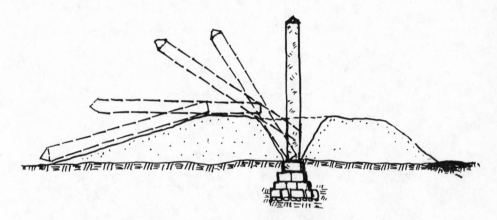

Figure 91: Erecting an Obelisk on a Foundation

Although no one knows for certain how Hatshepsut's Obelisk was raised, many have speculated how this feat of engineering might have been accomplished. Henri Cherier hypothesized that first a stone foundation and pedestal were built, then an earthen ramp was formed atop the pedestal. To place the obelisk, workers might have pulled the stone shaft to the top, using rollers and rails. When they had placed the butt end almost over the pedestal, they could begin excavating a pit. Slowly this would have rotated the obelisk to its vertical placement, and it would finally have rested on the pedestal.

Alternatively, a tunnel might have been excavated to the pedestal and the sand might have been removed, causing a pit down which workmen might have rotated the obelisk.

Whichever method the Egyptian workmen used, they had to use precise techniques to control such a huge object.

8
Prominent Engineers of History

Archimedes

This Greek engineer is best known for his discovery of the principle of flotation. A body in water, he said, has three natural states: Its density is equal to, less than, or greater than the surrounding liquid.

A body less dense than the liquid will float, and the portion of it that is submerged bears the same ratio to the whole volume of the body as its whole weight bears to that of an equal volume of water. If it is forcibly held below the surface, it will exert an upward force equal to the difference between its own weight and the liquid it displaces.

A body with a greater density than that of the liquid in which it is immersed will sink to the bottom, and if weighed while submerged, it will appear lighter than its true weight by an amount equal to the liquid it displaces.

Hero of Alexandria

A student of earlier technical writers such as Strato of Lampsacus, Ctesibus, and Archimedes, Hero discovered numerous physical principles that form the basis of much of modern engineering (Landels 1978, 9).

Hero's "Pneumatica" describes why a liquid flows through a syphon. His "Mechanica" discusses the pulley system, the wedge, and the lever. He also explains gear ratios and their implications, the parallelogram of forces, and block-and-tackle arrangements.

His "Metrica" deals with pure geometry, plane figures, triangles, and segments of circles. Hero also treated the art of surveying with a detailed description of the surveying instrument, sights, and water levels for adjustment of alignment. These instruments were used in conjunction with vertical poles and movable marker discs, to determine relative heights at various points. Hero also learned to use triangulation to measure the distance between two points not visible to each other and wrote technical treatises on construction of the heavy and hand catapults.

Some books attributed to Hero were (Gille 1966, 18):

"Pneumatica," on air and liquids
"Automatopoietike," on automatic devices
"Mechanica," on machinery
"Catoptrica," on the theory of mirrors
"Metrica," on pure geometry
"Dioptra," on surveying
"Geometrica," "Sereometrica," on solid geometry
"Belopoeca," on catapult construction
"Cheiroballistra," on the hand catapult

Vitruvius

You might call Vitruvius the founding father of architecture. His ten-book series, *De Architectura* ("On Architecture"), written for Roman Emperor Augustus, dealt with the rebuilding of Rome after its ruinous civil war of the first century B.C. In these books, Vitruvius laid the ground rules for the architect's work (Landels 1978, 36, 208).

The word *architect* comes from Greek words for the "supervisor" or "co-ordinator of various craftsmen." In Book I, Vitruvius explains the basic knowledge necessary for such a coordinator. He goes on to give a history of construction (Book II), from primitive mud huts to the types of buildings then being put up in Rome. He lists and describes materials of the time, such as sun-dried and kiln-baked brick, concrete made from pozzolana, marble, tufa, sandstone, and so on.

Vitruvius also covers site development, the design and decoration of temples, including the Doric, Ionic, and Corinthian orders, and public basilicas and baths (Books III to V). He describes the relationship between building design and climate (Book VI) and deals with both interior and exterior decorating (Book VII).

The last three books concentrate on engineering matters, including water supply, mechanical devices, optics, and even astronomy. He speaks of water pumps, waterwheels, catapults, and siege engines (even describing how to tune the skeins that power the catapults to the proper musical key) (De Camp 1963, 27).

Frontinus

Sextus Julius Frontinus specialized in water supply and engineering. Born in A.D. 35, he held his first political office in A.D. 70. In A.D. 97 he was responsible for the water-supply system for all of Rome, under Emperor Nerva. Frontinus was in charge of a total labor force of 700, including overseers, reservoir keepers, stonemasons, plasterers, and others. He personally inspected all water-supply projects throughout the Roman Empire (Landels 1978, 211).

Pliny

Gaius Plinius Secundus is known to history as Pliny the Elder, distinguishing him from his nephew, who became a Roman governor. Born in A.D. 23, he became chief of fleet under Emperor Vespasian in A.D. 69.

But Pliny was primarily a scholar. In thirty-seven rolls, or books, of papyrus, Pliny set forth his *Historia Naturalis* ("Natural History"), in which he claimed there were more than 20,000 important facts. After an introductory index of topics and sources (Book I), he went on to discuss astronomy (II); geography (III to VI); human and animal anatomy (VII to XI); botany and medical substances (XII to XXXII). The last five books were on minerals, including metals, stones, and the use of these materials in medicine, painting, and architecture.

9
Measurements, Mathematics, and Geometry

Although some early Hebrew inscriptions contain symbols that may be numerals, the Hebrew language has always counted on the letters of its alphabet to serve double duty as numbers. The first nine letters, *aleph* through *tet*, stand for the numbers one to nine. The tenth letter, *yod*, is ten, and the following letters are twenty, thirty, forty, and so on. The last four letters, *kof* to *tav*, are hundreds, 100 to 400. Of course, other numbers are formed by combinations of these letters. Eleven would be *yod aleph*, ten plus one. Two interesting exceptions are fifteen and sixteen, which were *not* written *yod he* and *yod vav*, as you would expect. Those were shortened forms of God's name. Instead, these were written as nine plus six (*tet vav*) and nine plus seven (*tet zayin*).

The Romans also used letters for numerals:

I	1
V	5
X	10
L	50
C	100
D	500
M	1000

The quantities in between were reached by adding or subtracting these amounts. By placing a smaller numeral before a larger one, you would be subtracting that amount. Thus IX is ten (X) minus one (I), or nine. By placing a smaller numeral after a larger one, you would add that amount. So VIII is five (V) plus three ones (III), or eight.

A bar over the letters indicated that the amount was to be multiplied by 1,000 ($\overline{\text{XIII}}$ is 13,000). There was no letter for zero.

Ancient man could also do more complicated calculations. In 1800 B.C. a Babylonian inscribed a clay tablet with a banking calculation, in cuneiform script. This writing was an algorithm, or step-by-step procedure for solving a specific problem, similar to modern computer programming. The scribe explained the procedure by presenting an example. Others could substitute the appropriate numbers for their own calculations. This procedure determined how many years and months it would take to double a certain quantity of grain (grain served as currency) at an annual interest rate of 20 percent.

Measurements

Ancient Egyptians measured distances in units equal to parts of the human body, as did many early cultures. The finger was generally the smallest unit. Four fingers equaled one hand. Three hands made a span (the distance from the tip of the little finger to the tip of the thumb, when the hand is stretched out). Two spans made a cubit (the length of the forearm).

As people began to trade with one another, these units needed standardization, to ensure fair practices. Thus the "royal cubit" became the standard in Egyptian, Sumerian, Assyrian, Greek, Persian, and Roman societies. Of course, the Sumerian king might have a longer forearm than the Egyptian pharaoh. Exact measurements might differ from nation to nation. But within each society, the standard was sure.

The Babylonians based their six-decimal system on a linear unit from astronomical calculations. This system was broadly accepted throughout the Middle East.

In Palestine, both the Egyptian and the Babylonian systems were used. The Hebrews, who had used the Babylonian system, introduced changes as a result of Egyptian influence. The Phoenicians and Persians added later changes. Thus the measurements mentioned in Scripture reflect varying standards.

The Persians had a distance measurement similar to the mile. It was 1,000 fathoms (2,000 yards) long. The word *mile* actually derives from the Latin *mille passuum,* "a thousand paces." The ancient Roman pace consisted of two steps, or about five feet. So 1,000 paces would equal about 5,000 feet, close to our modern statute mile of 5,280 feet. The Romans divided their mile into eight *stadia* of about 208 yards each (Brown 1962, 67–69).

LENGTH MEASUREMENTS

Cubit (two spans):
   17.49 inches (Old Testament), 18 inches (New Testament)
   20.6 inches, Egyptian (3000 B.C.)
   19.5 inches, Sumerian
   18.2 inches, Greek
Span (three handbreadths): 8.745 inches
Handbreadth (four fingers): 2.915 inches
Finger: .728 inches
Fathom: 72 inches, also man's height, four cubits, or the distance between
        fingertips, with arms outstretched.

Geometry and Surveying

The word *geometry* comes from two Greek words for "earth measurement." Geometry was originally developed by the Egyptians to determine the size of their fields and reset the boundaries of their farms each year after the Nile flooded. Thus, thousands of years ago, they already practiced an early form of surveying (Brown 1962, 68).

Throughout the ancient world, as nomadic tribes settled down and plotted out their ground, they needed accurate ways of measuring it. Even before the Israelites settled in Canaan, the Law spelled out the sacredness of boundary markers:

> Cursed is the man who removes his neighbor's boundary stone . . . .
> Deuteronomy 27:17 NIV

> Thou shalt not remove thy neighbour's landmark, which they of old time have set in thine inheritance, which thou shalt inherit in the land that the Lord thy God giveth thee to possess it.
> Deuteronomy 19:14 KJV

> Some remove the landmarks; they violently take away flocks, and feed thereof.
> Job 24:2 KJV

METHODS OF MEASUREMENT

Figure 92: Balance

Trade in precious metals and other materials required accurate standards of weight. This balance included a stand and a beam with suspended plates at each end; one held a standard, known weight, while the other held the unknown one.

Figure 93: Roman Steelyard

Ancient Romans used another form of the balance, including a hook, a beam marked with graduations, a fixed weight, and a suspended pan for the unknown weight.

Remove not the ancient landmark, which thy fathers have set.

Proverbs 22:28 KJV

Fair division of the land required some sort of surveying methods (Cornfeld 1981, 372; Brown 1962, 77).

Excavation of ancient cities indicates that those civilizations already knew a great deal about city planning, laying out the land, and so on. The Babylonians divided the circle into 360 degrees, a degree into sixty minutes, and a minute into sixty seconds.

But it took Greek mathematicians to begin developing the reasoning and logic needed to prove the truth of mathematical statements. Thales of Miletus

Figure 94: Determining the Weight of Floating Objects
Archimedes discovered that a floating body displaced its own weight in water.

(640–546 B.C.) began the study of lines and triangles. Pythagoras (580–500 B.C.) developed the theorem that bears his name, a way to determine the relative lengths of the sides of right triangles. Plato (427–347 B.C.) developed a logical method of forming a proof, and Aristotle (384–322 B.C.) noted the difference between axioms and postulates. Archimedes (287–212 B.C.) determined volumes and areas by methods now used in calculus.

Herodotus credited a king named Sesostris with the invention of geometry: "The king divided the land among all Egyptians so that to each he gave a quadrangle of equal size." But the Greek mathematician Euclid is generally considered the father of geometry, because he organized this form of mathematics into a single logical system. His book *The Elements* remains the foundational text in this field.

Geometric calculations gained importance as the complexity of life increased. Not only were more complex buildings being built, but increasingly warfare became waged with machines. A society's survival might depend not only on the strength of its warriors, but also on the accuracy of its mathematicians.

SURVEYING AND MAPMAKING

Figure 95: Ancient Land Surveyors at Work
A survey team included a notekeeper, rodman, and two chainmen, who set
points and marked distance with a measuring line (Brown 1962, 68, 69).

***Figure 96: Using a Bibical Plane
Table***
With this a surveyor prepared
maps or plats on the site.

Figure 97: Ancient Egyptian Rodman

10
Weapons, Fortresses, and Siege Engines

In the ancient world, as today, people continually developed new kinds of weapons, and each new measure in warfare was counteracted by something more powerful. As measure was followed by countermeasure and counter-countermeasure, both defensive and offensive weapons became increasingly sophisticated. War has always bred better machines.

Early Weapons

When David fought with Goliath, he described the Philistine's weapons: "You come to me with a sword and with a spear and with a javelin. . ."(1 Samuel 17:45 RSV). For personal combat, these were perhaps the most powerful weapons of his day.

The spears and javelins used by the Israelites when they entered Canaan had wooden shafts and metal points. Warriors used the spear for thrusting, while the smaller, lighter-weight javelin could also be thrown.

Early swords were generally short, straight, double edged, and used for stabbing in hand-to-hand combat. Judges 3:16 (TLB) describes Ehud's dagger: "Before he went on this journey he made himself a double-edged dagger eighteen inches long and hid it in his clothing, strapped against his right thigh."

During the Israelite conquest of the Promised Land, a sickle-shaped sword became common, for it could be easily swung from the chariot. But by the time David fought Goliath, the long sword, introduced to the area by the Sea Peoples, would have been in the Philistines' hands.

On the other hand, when he attacked the Philistine hero, David used a sling, something a shepherd might use against marauding wolves. The Bible expresses the surprise of David's victory over the well-armed giant:

> Reaching into his bag and taking out a stone, he slung it and struck the Philistine on the forehead. The stone sank into his forehead, and he fell facedown on the ground.
> So David triumphed over the Philistine with a sling and a stone; without a sword in his hand he struck down the Philistine and killed him.
>
> 1 Samuel 17:49, 50 NIV

Yet the sling eventually became a valuable battlefield weapon, since it gave a soldier a long-range threat in any terrain. This weapon was easy to make, and ammunition was plentiful. But consistently hitting one's target required a great deal of training and practice.

The sling consisted of a small patch of leather with two cords attached to the opposite edges. A stone was placed on the patch and the strings pulled, so the patch formed a small bag for the stone. The warrior swung this around his head several times, building velocity. At the proper time, he would release one cord, causing the stone to fly out at the same velocity along a tangent of its circular path. The slung stone achieved a far greater speed than one thrown by hand (Reader's Digest Editors *People* 1981, 41, 166).

The next great advance in weaponry, the bow and arrow, was probably the first man-designed weapon to concentrate energy. Most bows were made of an elastic wood or brass. The surface of the bow farthest from the string was called the back, the inner surface was the belly. The center, where the archer held the bow, was the grip, and the parts on either side of the grip were the arms or limbs. Bowstrings were made from the intestines of oxen or camels and were attached to the ends of the arms.

The arrow had three parts, each made of a different material to suit its function. The arrowhead was flint, bone, or metal, the hardest available material. The body, or shaft, was of reed or a light wood. This needed to be long, thin, hard, and strong, to direct the energy transmitted from the string on release. The tail, usually made of feathers, kept the arrow on course in a smooth and straight flight.

The archer placed the base of the arrow against the string and pulled the string as far as possible from the wood, creating maximum tension. This placed the bow under a bending stress, with tension on the back and compression on the belly. Since the wood had a high modulus of elasticity or rigidity, it conserved most of the energy expended in drawing the string back. When the archer suddenly released the string, the conserved (or potential) energy in the bow was also released. The bow returned almost instantly to its original shape, and the string propelled the arrow forward with great force and speed. The stabilizing effect of the heavy arrowhead and light tail feathers kept the arrow on course.

Some bows had an effective range of 300 to 400 yards, though an arrow could go twice as far with less accuracy. A good archer could shoot ten to fifteen arrows a minute. The crossbow, though smaller and easier to carry, lacked the range of the conventional bow and could fire only about two arrows per minute.

A bow's range depended on its size and shape, the pliability and toughness of the wood, and the archer's strength. The bigger the bow, the more energy it could store—and thus it could propel arrows farther. Yet large bows could be difficult to carry and operate.

Through time, warriors experimented with different shapes and materials for bows, trying to maximize the bow's tension without necessarily increasing its length. From the simple curved and triangular bows developed the single concave and double concave styles (*see* figure 98). At first, bows were fashioned from a single piece of wood. Later "composite" bows had several thin pieces of wood glued together; some had sections of bone in the belly.

Of course the archer needed a quiver to hold his arrows. Made of a light-

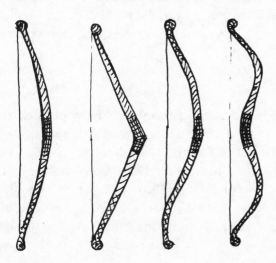

Figure 98: Shapes of Ancient Bows
From the simple curved and triangular bows (*left*), ancient warriors developed the single convex and double convex shapes (*right*), attempting to gain the greatest amount of tension while still maintaining a manageable size.

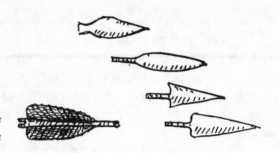

**Figure 99: Ancient Arrowheads
and Tail Feathers**

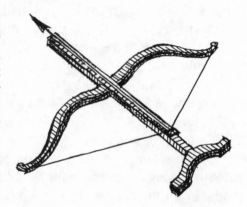

**Figure 100: Belly Weapon or Heron's
Gastraphete**
This early crossbow consisted of horn
and had a curved crosspiece at the butt
end for bracing against the chest or
stomach while arming (hence its name).
The main wood strut between the bow
and the butt ends contained a track,
bronze rack (or batten), pawns, and a
slide that held the dart or arrow (Ferrill,
1985, 171).

weight material, the quiver typically held twenty to thirty arrows and was carried
on the back or over the shoulder, so the archer could keep both hands free.

The bow and arrow required both strength and skill, but it substantially
increased an army's range of attack. Not surprisingly, the Bible uses the bow as
a symbol of strength. In Hannah's song, the Lord reigns, and "the bows of the
warriors are broken . . ." (1 Samuel 2:4 NIV).

Armor

Early warriors wore helmets and body armor and carried shields for protection.
Archers generally carried smaller shields, if any, while spearmen and swordsmen
carried large shields that protected most of their bodies.

Early helmets, made of brass, were mainly reserved for officers, nobles, and
the more important warriors. Some also wore long coats of mail, extending to the
knees or ankles, while common soldiers generally wore only leather jerkins,
protecting their upper bodies. The Bible also mentions brass greaves, which
protected the legs (1 Samuel 17:4–7).

WARRIORS OF THE BIBLICAL AGE

Figure 101: Ancient Warriors
Biblical soldiers, from left to right, include: Assyrian, Babylonian, Persian, Judean, and Roman (Reader's Digest Editors *People* 1981, 13, 15, 97; Reader's Digest Editors *Atlas* 1981, 3, 5, 15).

Figure 102: Headgear of an Assyrian Soldier

Figure 103: Assyrian King Sennacherib With Helmet and Armor
Armor scales with fastenings and lacings protected the upper body.

Figure 104: A Canaanite Warrior

Figure 105: Headgear of a Philistine Warrior

***Figure 106: Helmet of a Roman
Soldier***

Figure 107: Headgear of a Roman Legionnaire

Figure 108: Full Battle Gear of a Roman Infantryman
(Cornfeld 1981, 45, 179).
Josephus described the
infantryman's equipment as
follows: "The infantry was armed
with a coat of mail and a helmet,
and carried a sword at each side;
that on the left side is by far
longer of the two, the dagger being
no longer than a span in length.
The general's bodyguard of picked
infantry carry a javelin and long
shield, together with a saw,
basket, ax and pick, as well as a
strap, a chain and three days'
ration so that an infantryman is
about as heavily loaded as a pack
mule" (Reader's Digest Editors
Atlas 1981, 17).

***Figure 109: Fully Equipped
Roman Trooper***
The spear and short sword are
held on the right side. For
personal protection the soldier
wears a crested helmet, leather
breastplate, and greaves (Reader's
Digest Editors *Atlas* 1981, 17.)

Figure 110: An Egyptian Archer

Fortresses

Two major factors determined the sites of fortresses in ancient times—strategic position and water sources. These factors often conflicted with each other: Hilltop sites were easier to defend, but water sources usually lay in the valleys. Thus many fortresses were built on hills *near* well-watered valleys, with the fortress walls enclosing a valley water source. On occasion, a deep cistern or pit would be dug inside the walls and connected by a tunnel to the stream or spring outside the walls. This was Hezekiah's solution to Jerusalem's water problem (2 Kings 20:20).

Walls served both defensive and offensive functions. Obviously, they prevented enemy troops from overrunning a fortress, but they also gave defenders a position from which to attack the enemy. The Bible describes the war preparations of King Uzziah of Judah (791–739 B.C.): "In Jerusalem he made machines designed by skillful men for use on the towers and on the corner defenses to shoot arrows and hurl large stones. . . " (2 Chronicles 26:15 NIV).

Walls needed to be thick and high to prevent battering and scaling—and deep enough to keep the enemy from tunneling under them. They also needed the strength to withstand catapulted rocks. Moats dug around the walls might keep attackers away. Sloping walls, or glacis, would deflect battering rams' impact and would enable defenders to roll large rocks down upon an onrushing enemy. The top of the wall had to give defenders a good vantage point to see the attackers and to fire on them. Fortress-wall design could leave no "dead spots," concealed corners that an enemy might scale unobserved. Thus many fortresses had irregularly shaped walls, crenelated parapets, special towers, and bastions for the defending archers and spearmen.

The gate of a fortress was its most vulnerable part. It needed to give friends access and deny entrance to foes. Thus ancient military engineers designed gates carefully, usually placing them at the steepest part of the hill. An access road had to wind back and forth, maintaining a practical slope for those ascending, but keeping unwanted enemies from rushing the gate. Where possible, the road reached the gate with the fortress on the right side. This exposed the attacker's right side to fire from the wall. Since a warrior generally held his shield on his left arm, the attacking soldier was especially vulnerable on his right.

In some cases two gates protected a city or fortress. These were usually situated so that if attackers managed to penetrate the outer gate, they would have to approach the main gate with their right sides exposed.

Doors at these entryways were generally made of metal-clad wood, to keep attackers from setting them afire. Their opening had to be wide enough to accommodate chariots and wagons, so two swinging doors typically covered the space, meeting at the center. A heavy beam, lying across the doors and set in recessed sockets of the stone walls, would lock the doors. Sometimes the sockets would be lined with bronze to reduce wear and tear.

The gate was also protected by towers on either side. From these vantage points, defenders could fire arrows and spears or pour burning oil on anyone attempting to ram the doors or smash their hinges. Some fortresses even had a roof with holes constructed over the gate area, which allowed the defending soldiers to fire directly down upon the attackers' heads. In essence, there were often subfortresses built around a fortress gate, since that was generally the structure's most vulnerable point (Ussishkin 1988, 42–47).

The Fortress at Jericho

Dating to the Middle Bronze period (about 2000 B.C.), this fortification included a glacis of beaten earth with a brick wall at the top. Over time, two more glacis were added on top of the first one. The last of these, a stone embankment, was built about 1750 B.C. and destroyed about 1550 B.C. (*see* figure 112 for a typical glacis and moat) (Paul 1973, 84, 85).

Figure 111: Through Section of an Ancient Protective Stone Wall

To start, the builders needed solid ground upon which they could place a stone foundation. On top of this they could begin building a wall (*right*) made of stone masonry and measuring nine feet high by four feet thick. Next they covered the outer face of the wall with small stones and mortar. To finish it they packed a layer of earth on top of the mortar and in turn covered that with a sixteen-inch-thick layer of pebbles and mortar (Mazar 1975, 153).

Figure 112: A Defensive Wall With Glacis and Moat

To discourage attacks by tunneling and siege machines, the builders of this more sophisticated wall erected a glacis and excavated a moat that allowed the defenders to counteract enemies with large stones, which they could roll down at attacking troops.

The builders began by cutting a bench into the side of the hill. On top of a rubble foundation they built a vertical wall of cut stone. To form the glacis, they sloped the front of the wall and built it up with alternate layers of beaten earth and stones. Beneath the slope they excavated a moat.

Saul's Fortress-Palace

The double wall and broad towers of Saul's fortress-palace at Gibeah made it the strongest Israelite citadel of its time. The space between the inner and outer walls was partitioned into chambers and used for storage. But this space could also be packed with rubble to make one solid wall, thwarting any battering-ram attacks. Crenelated parapets gave defenders narrow gaps through which archers could fire arrows, while providing ample protection from the missiles of the enemy.

Solomon's Fortress at Megiddo

Megiddo lay at a crucial crossroads, where the Jezreel Valley met the famed Via Maris. Numerous armies have marched through the mountain passes in that area.

Such a strategic location demanded a solid defense, and King Solomon provided it. The city gates boasted a brilliant design. Guard towers looked down on the entryway to the city, which could only be reached by a steep stairway or an exposed ramp. The outer gates had two pairs of heavy wooden doors, which opened into an enclosed courtyard with high walls. This led to the main gate, which had four sets of heavy doors. Between these were small spaces, where soldiers could be posted to surprise intruders.

The Fortresses of David and Solomon at Taanach

Earlier structures of brick and stone had fallen at Taanach, a few miles south of Megiddo, but Solomon used some of those materials to build a network of garrison forts. The west fort, sixty-two feet by seventy feet, was built with large boulders (cyclopean masonry). The walls, four feet thick, sat on a foundation that went down thirteen feet. A thick layer of lime and plaster coated the floor. The fort included nine or ten chambers, most likely used for storage. Archaeologists surmise that the structure was intended merely as a framework for a more impressive building or fighting platform to be built on top.

The northeast tower was built of limestone, finely dressed, squared off, and laid in regular courses. It included openings through which defenders could shoot (Reader's Digest Editors *People* 1981, 201).

Herod's Fortress at Masada

Herod had a way of building fortresses at the most inaccessible sites. A prime example was Masada, on a steep mountain on the west shore of the Dead Sea. When the Romans invaded Israel to put down the Jewish rebellion of A.D. 67–70, they found Masada the hardest to conquer. They besieged the fortress, but the mountaintop was self-sufficient with food and water. Eventually the Romans built a circumvalation, or siege wall, around the mountain. After two years, they finally worked their way to the top—only to find that the Jewish defenders had committed suicide rather than be captured.

Machines of War

Well-built fortresses gave defenders the advantage. The man at the top of the wall could shoot an arrow, throw a javelin, drop a boulder, or pour burning oil or molten lead on an attacker. Thus, attacking armies needed countermeasures, siege machines or war engines that could successfully overcome the defenders' advantage. This meant employment for military engineers, specialists who designed, built, and operated such machines.

Often engineers had to improvise. Ancient armies generally would not carry bulky catapults with them—that would slow them down. Instead, the engineers packed the skeins, slings, and metal fittings they needed and pulled together the other materials on the spot. As an army prepared to besiege a city, soldiers would cut down trees in the area and build their war machines there (Cornfeld 1981, 373).

The battering ram, depicted on reliefs of the Assyrian king Ashurnasirpal (883–859 B.C.), was one of the earliest war machines. At first just a large wooden beam carried by soldiers, it developed into sort of a combination of the modern tank and bulldozer. Engineers developed a protective housing for the beam. This wooden frame, sometimes mounted on wheels, might have rectangular wicket shields on the sides and a domed metal turret on top. Inside the turret, ropes held the battering ram in place. The entire machine might be five meters long and about eight meters high. Eventually, military engineers fixed an axlike blade or metal point on the front of the ram. This device, often called a bore or pike, would be driven into joints between the stones of the wall. Once the blade or point went deep enough, it could be moved from side to side, loosening the stones and causing the wall to collapse. When attacking the gates of a fortress, pikes, swords, and spears were used to pry the gates loose, tearing out the hinges. Of course, this all had to be done with shields or some sort of protective structure overhead.

The prophet Ezekiel foretold the use of battering rams and other war machines in the Babylonian conquest of Jerusalem: "The king of Babylon . . . is to set up battering rams, to give the command to slaughter, to sound the battle cry, to set battering rams against the gates, to build a ramp and to erect siege works" (Ezekiel 21:21, 22 NIV).

Classic writings tell of the Phoenician engineers who introduced a battering ram suspended by chains from the roof of a wheeled shed. This was called a ram tortoise.

Assyrian military engineers excelled in siege warfare. One of their methods used the weight of a fortress wall against itself. Workers—"sappers," they called them—would dig a hole under a wall, temporarily shoring it up with wooden supports. But then they would set fire to the wood, and the wall, left unsupported, would collapse.

Though not exactly a machine, one method used by the Syrians against the Maccabees in the second century B.C. is strangely akin to the modern tank. They built wooden towers on the backs of elephants—thirty-two of them—gave the animals wine, and irked them into charging the enemy. Eleazar, brother of Judas Maccabeus, killed the largest elephant, but he died when it fell on him (1 Maccabees 6:28–47, Apocrypha).

The Catapult

The earliest catapults shot large arrows, up to six feet long. These machines were essentially massive crossbows, mounted on pedestals.

Figure 113: Ancient Egyptian Sappers

This painting, found on a tomb wall, shows three Egyptian combat engineers in a portable hut. A reed and basketwork frame covered with hide protected these sappers, who dislodged the stones from a fortress wall (Yadin 1963, 17).

Figure 114: Ancient Assyrian Battering Ram and Archers

Military engineers developed a rather sophisticated mechanism for cocking and releasing these huge bows. A wooden beam (the syrinx or trough) was mounted on the pedestal at an angle. This had a large groove in its upper surface. At the top of the trough lay the crosspiece of the bow, fastened by brackets. Within the groove of the trough beam, a smaller beam (the diostra, projector, or slide) moved back and forth. This had a narrow groove in its upper surface and a piece of metal, or trigger, at the lower end. It had a hook to engage the bowstring and some kind of knob, handle, or lanyard to release it.

A windlass at the lower end of the trough pulled the slide piece down the groove of the trough, stretching the bowstring to the desired point. The metal trigger on the slide also had two pawls (or dogs), which clicked over the teeth of two bronze racks that ran the length of the trough's groove. These kept the slide from slipping back during the winding of the windlass.

When the bowstring was stretched tight enough (depending on the desired distance of the shot), the windlass was stopped; the teeth held the slide (and bowstring) in place, and a dart was loaded into the slide's narrow groove. Then the bowstring would be disengaged from the hook by means of the trigger mechanism and would snap forward, sending the dart on its destructive journey.

Figure 115: Ancient Babylonian Combat Engine
(Ferrill 1985, 75; Yadin 1963, 18).

Then the slide would have to be pulled up the trough, by hand or windlass, to engage the bowstring again.

The dart-throwing catapult more than doubled the effective range of an army. The most powerful longbows might accurately shoot 300 to 400 yards, but the catapult was effective at 600 to 800 yards (Landels 1978, 99).

Later engineers—possibly the Phoenicians—developed the stone-throwing catapult, attaching a pouch or strap in the middle of the bowstring to hold the stone. History first mentions this weapon in connection with Alexander's siege of Tyre (332 B.C.). Its range depended on the weight of the projectile (10 to 180 pounds). A 60-pound stone might fly 200 yards. The dart throwers' greater range made them more effective against enemy soldiers, but the force of the stone throwers' heavy missiles made them better against structures, such as ships or siege towers. One problem with the stone thrower was that, by attacking, you might also arm the enemy. If they had catapults, they could fling your stones right back. Some armies began catapulting bricks, which usually broke on impact, so they could not be shot back.

The torsion device took the catapult a step farther. This used a braided rope of horsehair or human hair for tension, rather than a wooden bow. A massive frame surrounded the trough, and the rope was threaded through the sides of this frame until two strong skeins were built up. A pair of rigid throwing arms, thrust through these skeins, took the place of the arms of a solid bow.

The torsion catapult stored more energy than the flexion catapult and thus could throw heavier missiles, and throw them farther. Yet it also proved harder to manage. The skeins would go slack in wet weather or with long use and needed regular tightening.

The ballistae and some other forms of the catapult used a skein of animal tendons to store and release energy. The energy stored in 36 pounds of tendons could throw a 90-pound missile 400 meters. Modern-day engineers estimate that it would take 700 pounds of steel spring to match this.

Tendons became the material of choice in Roman catapults. Though some writers allowed that human hair might be substituted, the military writer Vegetius insisted, ". . . ballistae are no use unless powered by sinew-rope." Yet despite the amazing amount of energy that could be stored by tendon material, those ancient weapons were limited by the fact that the artilleryman had to crank in all the energy to pull the material taut (Vogel 1986, 68).

The onager siege machine, another development in the catapult, had a single arm connected by a torsion skein. When stretched and released, the skein propelled the arm in a vertical plane against a padded stop. At the end of the arm, a sling or spoon held the projectile. (*See* figure 116.)

The onager gave way to the trebuchet, or counterweight catapult, an enormous sling placed at the end of a jib, which carried a counterweight at its other end. The jib pivoted on two pins, and a large stand supported the whole device. Since the jib was vertical and the counterweight very heavy, lowering the jib or raising the counterweight took great effort. Therefore the engineers constructed springs consisting of pieces of flexible wood and assembled in a T shape, to ease the load. A cable with return pulleys and a turning device, which led to the end of a strong bar fixed to a winch, stretched the springs. When they were taut or the bar turned back, it required a counterweight of 1,300 cubic feet of earth to hold the machine steady.

Because it did not rely on the tension of a skein, the trebuchet worked well even in wet weather. By varying the size of the counterweight and its distance from the fulcrum, engineers could adjust the weapon's range. Some simple forms of the trebuchet used the force of soldiers pulling on ropes to propel the missile (Morgan 1960, 285).

Naval Artillery

The machinery of warfare developed on sea as well as on land. In 313 B.C., Demetrius, who had worked with siege engines on land, put catapults and ballistae on galley ships. With dart-throwing catapults on the prow, his vessels

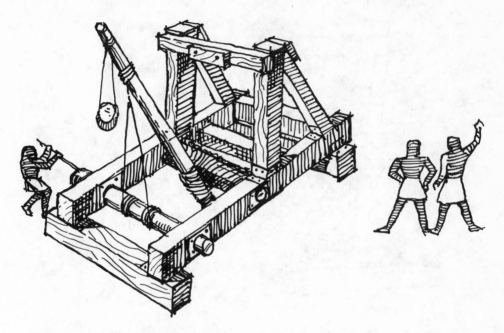

Figure 116: An Onager Catapult

A war machine that worked particularly effectively against oncoming armies was the onager catapult, or one-armed torsion stone thrower.

The sturdy, bolted frame held a heavy, padded stop that halted the arc of the wooden throwing arm. A torsion skein of rope or women's hair would conserve the energy provided by a soldier, who manually turned the winch or windlass behind the arm.

From the wooden arm hung the sling that held the ballista stone. By turning the winch, the soldier pulled the arm back, tightening the skein so he could fit the stone into the sling. With a trigger mechanism, he released the stored energy in the skein, sending the arm swinging rapidly against the stop and throwing the stone forward.

Because this machine could throw such enormous projectiles, defenders could crush men, horses, and enemy machines that came before their walls. Naming it an *onager*, after the wild ass, reflected the catapult's kick.

Figure 117: A Torsion Dart Thrower
Also known as quick firers or scorpions, these catapults were generally
mounted on mule-drawn carts. Each century of a Roman legion would have
one, and it would be manned by two soldiers, turning the windlass and
loading the darts. The weapon could fire a barrage of spears about 300
meters.

    As Vitruvius described its use: "The catapult operated the same way
as the later cross-bow. The pair of vertical coil chambers was at the front,
the bow being drawn back by a windlass to the required limit of tension.
The bow was released by a trigger and the bolt shot along a trough with
an aperture at the front" (Mazar 1975, 90).

could launch preliminary barrages against enemy craft long before they came
within bowshot. Through time, naval catapults used larger and heavier ammunition, such as stones, which could do considerable damage to enemy ships, and
"Greek fire," stones covered with cloth soaked in naphtha, crude oil, sulfur, and
pitch (Ferrill 1985, 85).

    Fire proved to be a crucial weapon in naval combat. In the second century

Figure 118: A Shipboard Catapult
(Mansir 1981, 46).

B.C., the Rhodians won sea battles by hanging pots with Greek fire in front of their galleys.

Later warriors learned that saltpeter, added to the Greek-fire recipe, caused spontaneous combustion. This led to two more types of incendiary weapons. One was a catapult that tossed clay jars filled with a self-igniting mixture. These jars broke on impact, creating an explosion. The other weapon was actually a rocket launcher. A reed would be filled with the Greek-fire mixture, stoppered, and placed in a bronze tube on deck. The tube would be aimed and a fuse lit. When the reed ignited, it released gases that propelled it out of the tube and in a flaming streak toward the target. In close combat between vessels, the Greeks used another form of this weapon. A large tube of wood lined with bronze would be set on the foredeck and connected to an air pump. The tube would be filled with Greek fire, then lit. As the sailors worked the pump, a shaft of flame shot outward, as in a modern flame thrower.

Pericles used a similar weapon in land warfare at Samos about 441 B.C., when he commanded his soldiers to set fire to wooden stockade walls:

> They . . . hollowed out a great beam . . . and suspended a vessel by chains at the end of the beam; the iron mouth of the bellows directed downwards into the vessel attached to the beam, of which a great part was itself overlaid with iron. The machine they brought up from a distance on carts to various points of the rampart . . . and when it was quite near the wall they applied a large bellows to their own end of the beam, and blew through it, the

blast, prevented from escaping, passed into the vessel which contained burning coals and sulfur and pitch; these made a huge flame, and set fire to the rampart, so that no one could remain upon it (De Camp 1963, 101).

In Roman conflicts of 38 B.C., Agrippa used a variation of the dart-throwing catapult in his battles against the lighter and faster ships of Sextus. His catapult shot an iron rod, tipped with a grapnel and attached to a line. Shot at an enemy ship, it would sink into the vessel, and sailors could pull the line, bringing the two ships close together. Agrippa's troops could then board the enemy ships and defeat them in hand-to-hand combat. The catapult gave Agrippa much greater range than the hand-thrown grapnel, a necessity against the swift craft of Sextus.

How These Machines Worked

Each of these machines of war worked according to certain properties of physics, making use of predictable scientific laws. The engineer merely harnessed them to achieve his goals.

The battering ram simply used the weight of the massive beam against the wall or structure being assaulted. When it swung, the kinetic energy of the moving ram converted suddenly into explosive energy, shattering the stone of the wall. By increasing the ram's weight and lengthening the ropes suspending the ram, engineers could provide the swinging ram with greater kinetic energy. In addition, they could increase the manpower behind the ram.

The catapult used the potential energy stored in the stretched skein. When released, the skein transferred this energy to the arm, which swung upward in an arc until it was stopped suddenly by the heavy frame. The projectile—the stone or arrow being thrown—would continue in a direction tangential to the arc of the arm's movement. Its initial velocity would equal the speed of the arm just before it stopped. This velocity depended on the degree of tension in the rope and how far the arm had to move before being stopped by the frame.

Ancient military engineers learned to adjust the projectile's direction and speed (and therefore its range) by setting the catapult at different angles and winding the windlass more or less. They had to be aware not only of the laws governing the act of sending the missile into the air, but also of the laws of gravity, which pulled the missile back to earth—and to its target.

The initial velocity of the missile can be divided into vertical and horizontal components—that is, at this particular angle, how fast is it going *up* and how fast

Figure 119: A Grappling Catapult

The Romans mounted this offensive weapon, consisting of a pedestal, bow, bowstring, and iron grappling dart, on the gunwales of their men-of-war. When the soldier cut the line that held the bow taut, the dart shot across to the deck of the enemy ship and lodged there. The attackers used the rope attached near the dart's head to pull the craft close for boarding (Mansir 1981, 47).

is it moving *across* the land toward the target? Next one determines time of flight by simply taking the vertical velocity and subtracting the acceleration due to gravity. This would be similar to shooting an object straight up and seeing how long it took to fall back to earth—at some point the downward pull of gravity will overcome the upward power of the initial shot, and the missile will fall. Then, when the time of flight is determined, one can figure from the horizontal velocity just how far the missile will go across the terrain in that time.

Catapult operators had two main variables as they aimed their weapons: the angle of the throwing arm at impact and the amount of tension in the skein. Once they knew the speed at which so many number of turns on the windlass would propel a stone of a certain weight, their calculation of the angle would give them a good idea of where the stone would fall. To increase that velocity, they would tighten the skein more. To change the relation between vertical velocity and horizontal velocity, they would change the angle. In modern artillery calculations, air resistance and other factors are also considered, but ancient artillery-men probably compensated for these factors by trial and error.

The Roman Assault on Jerusalem

One of the best examples of the use of military engineering in ancient warfare comes from Jewish history: the Roman conquest of Jerusalem in A.D. 70.

The Jews had been under Roman rule since 64 B.C. Although they prospered economically during this time, the Jews chafed under the insensitive treatment of puppet kings like the Herods and later governors like Pontius Pilate. In A.D. 43 Emperor Caligula even tried to install a statue of himself in the Jerusalem Temple.

The spark of rebellion came in A.D. 66, when the Greek population of Caesarea turned on the Jews who lived there. Josephus says they slaughtered 20,000 Jews in one hour. The Roman garrison stationed in Caesarea did nothing to stop the bloodbath.

The Jews of Jerusalem retaliated with a successful attack on the Roman garrison in their city. Then the Roman governor of Syria marched toward Jerusalem with the Twelfth Roman legion, to put down the uprising. But the Jews ambushed this army in a narrow pass at Beth-horon. Soundly defeated in the battles of the day, the Romans beat a hasty retreat at night, leaving behind their heavy equipment, including catapults and slings.

The thrill of victory didn't last long for the Jewish rebels, as Rome's massive army mobilized to quash the revolt. Thirty legions (of about 5,000 men each), throughout the empire, were alerted. Ten legions were stationed within a two-month march of Jerusalem. They came with their war machines: the swinging rams, the catapults, the rapid-fire dart throwers. By contrast, the Jews relied on guerrilla warfare: hand-to-hand combat and a thorough knowledge of their rocky terrain.

By A.D. 70, General Vespasian, conqueror of Britain, had bottled up the rebels in Jerusalem. He returned in glory to Rome, where he was named emperor, leaving the siege of Jerusalem to his son, Titus.

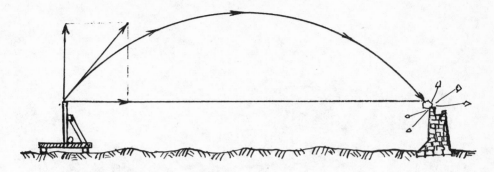

Figure 120: Catapult Trajectory
To figure the most effective way to hit his target, the ancient artilleryman
had to know the physics behind his machine. When the catapult's throwing
arm hit the stop, the projectile would fly upward, continuing the arc the
swinging arm had begun. The stone would continue at the velocity at
which it had begun until the force of gravity pulled it down, sending it in a
parabolic curve. As it descended it would strike the ground or target. To
adjust the missile's flight, the engineer would adjust the skein that
provided the energy or alter the place at which the arm hit the stop.

The Tenth Legion approached Jerusalem and set up camp on the Mount of
Olives, which overlooked the city from the east. The Jews attacked this encamp-
ment and were rebuffed by Roman reinforcements. In response, Titus cleared the
gardens just outside the walls—to prevent any more sneak attacks—and set up
his assault force on the northwest side of the city.

About 150 yards northwest of what is now Jaffa Gate, the Romans began
assaulting the wall of the city with three battering rams. They also wheeled three
assault towers into place adjacent to the walls. These towered over the walls,
enabling Roman archers to fire on any Jewish defenders who might try to fire
down upon those operating the rams. In addition, the ram force was protected
with wooden shields covered with hides, to deflect flaming arrows and burning
oil. Roman archers, bolt firers, and slingers also maintained a steady barrage at
the parapets and any openings in the wall. Further back, Roman slings and
catapults added to the assault.

Finally the rams broke through the wall, and the Romans entered, forcing
the Jews to retreat to the second wall. The Romans then began to assault the wall
southwest of what is now the Damascus Gate. Five days of battering broke this
wall, too, and 1,500 legionnaires moved into the Jerusalem marketplace. Sud-

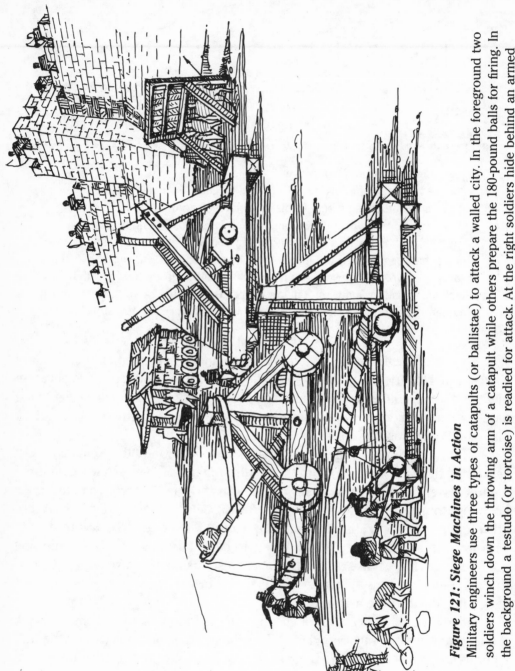

Figure 121: Siege Machines in Action
Military engineers use three types of catapults (or ballistae) to attack a walled city. In the foreground two soldiers winch down the throwing arm of a catapult while others prepare the 180-pound balls for firing. In the background a testudo (or tortoise) is readied for attack. At the right soldiers hide behind an armed approach shield (Yadin 1963, 16–24, 314, 437; Rabinovich 1971, 50–55; De Camp 1963, 181).

denly the soldiers found themselves in the winding alleys of an unfamiliar city. This was the Jews' turf, and the one-to-one combat on the narrow streets worked to their advantage. The Romans had to pull back again.

The Jews quickly rebuilt the wall, but in five days the Romans had broken through again. This time the soldiers burned the marketplace and forced the defenders behind their inner wall. The Romans sent legions to attack the Upper City and the Antonia Fortress, while their engineers built a huge assault tower.

But the Romans weren't the only ones using military engineering. The Jews had about 150 catapults and other war engines, captured from the Twelfth Legion, four years earlier. They had trained themselves to operate these and bombarded their attackers with some success. They also employed an ingenious plan to build tunnels under the areas where the Romans would place their assault towers. As they dug, they shored up the tunnel with wood struts covered with pitch. Then they filled the shaft with combustible material. When the Romans rolled their heavy towers up to the wall, the Jews set fire to the tunnel, burning the wooden struts. The ground gave way, and the assault towers toppled.

Realizing how important the Romans' machines and towers were, the Jews launched a series of forays against those weapons themselves. Because the Romans were running out of building materials for new machines, they had to go farther and farther for timber. But in essence, the Jews just bought time. They held the fort against the most powerful army the world had known, and the Romans kept battering away at the walls.

Eventually, the wall over the tunnels the Jews had dug collapsed. The Romans streamed in, engaging the defenders in hand-to-hand combat. After about a month, the Romans broke into the Temple area. Josephus says General Titus wanted to spare the Temple, but his soldiers ran out of control. They burned the Temple and eventually pushed its rubble over the eastern retaining wall.

Figure 122: An Assault on a Fortress

As defenders shoot their bows and arrows and lose spears against them,
the attacking soldiers make use of a wheeled siege tower, covered in the
front with raw hides and fitted with a gangplank (*at top*) and a battering
ram (*at the base*). Meanwhile, to each side, more soldiers climb a scaling
ladder that reaches the top of the wall (Cornfeld 1982, 39, 231).

11
Water Supply

People need water to drink, to wash in, to water crops, and to cook with. Societies have always depended on their springs, wells, lakes, or rivers. In ancient times, cities grew around water sources. But as they grew, they needed better water management. Greater population meant greater demand for water—and potentially greater pollution of the water supply. The health—and sometimes survival—of the society was at stake.

Great civilizations learned to manage their water well. They developed cisterns, aqueducts, and piping systems for the inflow of fresh water and the disposal of waste.

Approximately 5,500 years ago, in Egypt, palace apartments had copper water pipes. The designers of the Great Pyramid also installed copper piping for the royal bathing pool of Pharaoh Cheops. To the northeast, the Babylonians dug a network of canals connected to large underground drainage sewers made of brick. In the eastern Mediterranean, homes in Crete, dating to about 1,000 B.C., had bathtubs and toilets made of hard pottery.

The Romans are often credited with the invention of hydraulic engineering, but even they borrowed from their Etruscan predecessors. Rome's Cloaca Maxima, which still carries water to the Tiber River, was built by an Etruscan king about 400 B.C.

Yet the Romans *did* bring the art of sanitation to a higher level, building massive aqueducts and extensive underground sewer systems, public and private baths. They used lead and bronze piping and marble fixtures with fittings of gold and silver. Much of this was plundered when the Vandals and Goths invaded Rome. Succeeding centuries saw a great neglect of water management, resulting in widespread disease.

Water Conduits

Many Mediterranean countries used pipes to conduct water. Much of this piping was made from lead, which was plentiful in Spanish mines. (Vitruvius would be one of the first to warn of this metal's health hazards.)

In the foundations and retaining walls of Jerusalem, one can still see the hollowed stone blocks that ancient peoples used as piping. These blocks were squared, hollowed out, and grooved to fit together. The resulting pipes would be laid out on the ground, transporting water to homes and fields throughout the Judean hills.

Cisterns

The inhabitants of Jerusalem dug cisterns for water storage as early as 2000 B.C. The limestone of this region suits that purpose well—coated with lime plaster, it held water very effectively. In some cases, houses were built from the stone that was cut out of the ground during cistern construction.

Many such household reservoirs still exist in Jerusalem's rocky ground. Cut sixteen to twenty-three feet deep, they are generally bottle shaped. That is, the base is a cube or cylinder with vertical walls, but it narrows near the top to an opening about three feet in diameter. Many cisterns include depressions in the bottom to catch mud and sand sediment, so that a jar lowered into the cistern will bring back only clear water.

The Bible indicates that cisterns were already built in Palestine before the Israelites settled there. The Lord told them that the Promised Land included "houses full of all good things, which you did not fill, and cisterns hewn out, which you did not hew . . ." (Deuteronomy 6:11 RSV). King Uzziah's building projects included public cisterns built in the desert, to enable herdsmen to water their animals. As part of the propaganda of Assyria's threat to Jerusalem, King Sennacherib promised the Jews: ". . . Make peace with me and come out to

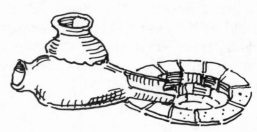

Figure 123: A Potsherd Drain
Such a drain directed water collected from a flat roof into the mouth of a cistern (Canaan 1954, 18).

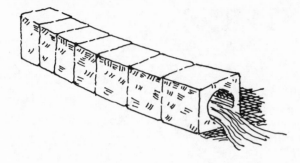

Figure 124: A Hollowed Stone Pipe

me. Then every one of you will eat from his own vine and fig tree and drink water from his own cistern" (2 Kings 18:31 NIV).

The land of Israel can be very dry, but it also has regular times of rain. Thus it became essential to make the best use of that rainwater when it came. For this purpose, the flat roofs of houses were plastered, curbed, and slightly sloped. Rain would collect in a corner of the roof and descend through clay pipes to the cistern or a large urn. Courtyards were often paved with stone and similarly curbed and sloped so that rainwater would flow toward the underground reservoir.

Water-Supply Sites

Ophel
The history of the settlement of Jerusalem demonstrates the importance of water supply and storage. The first inhabitants lived on Ophel, the lower eastern ridge of the city. This was quite vulnerable to attack, but it was close to the water supply—the springs in the Kidron Valley. After the development of lime mortar, about 1000 B.C., cisterns could be made quite waterproof, but the first residents of Jerusalem still used mud mortar, which allowed a great deal of seepage. Thus they had to live close to running water. After lime mortar came into use, residents depended less on the Kidron springs. They could settle the higher ridges and live off the rainwater from their watertight cisterns.

The Temple Mount
Under the Temple Mount's courtyard lie at least thirty-seven reservoirs. The largest, *Bahr* ("the lake"), is 40 feet deep, 246 yards in circumference, with a

CISTERNS

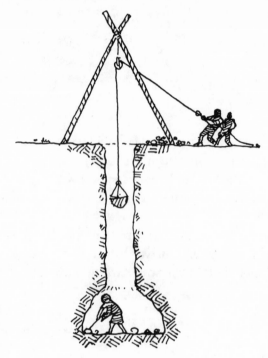

Figure 125: Excavation of a Cistern
(Canaan 1974, 62).

capacity of 13,600 cubic yards (12,000 cubic meters). Some say it was supplied by the aqueduct from the Solomon Pools, near Bethlehem.

Hezekiah's Tunnel

As Jerusalem became larger and more important it needed a safer water supply. Under normal circumstances, the cisterns would suffice, supplemented by occasional trips to the Gihon spring outside the walls, in the Kidron Valley. But in times of siege, the city would be cut off from this spring. When the Assyrians ran rampant through the Middle East, Hezekiah decided to prepare for a possible siege by digging a tunnel from the spring to a reservoir within the city walls. Using hammers and chisels, Hezekiah's workmen carved their way through 1,749 feet of solid limestone, removing some 850 cubic yards of rock. Two teams

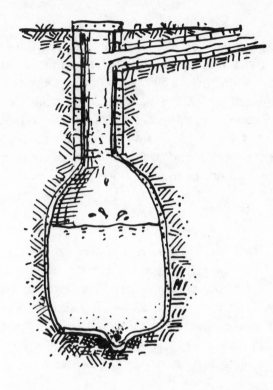

Figure 126: An Ancient Cistern
To build this reservoir, the solid
rock had to be excavated, to form
a basin that held the water, a shaft
that reached the surface, and a
drain that carried the water to the
shaft. Once the rock was removed,
workers applied lime plaster to the
cistern and used stone and cement
to line the shaft. Finally they
sealed the top of the shaft with a
protective stone cap.

apparently worked simultaneously, one digging from the spring, the other from
the reservoir. It took an estimated 200 days—and the tunnel twists and winds
through the rock—but the two teams eventually met.

In 1800, a schoolboy exploring the tunnel found an inscription announcing
the completion of the project. Though only the latter part of the inscription
remains, and parts of it are unclear, it still gives a feeling of the joy those
workmen must have experienced:

> [. . . When] (the tunnel) was driven through. And this was the way in
> which it was cut through:—While [. . .] (were) still [. . .] axe(s), each man
> toward his fellow, and while there was still three cubits to cut through, [there
> was heard] the voice of a man calling to his fellow, for there was *an overlap*
> in the rock at the right [and on the left]. And when the tunnel was driven
> through, the quarrymen hewed (the rock), each man toward his fellow, axe
> against axe; and the water flowed from the spring toward the reservoir for
> 1,200 cubits, and the height of the rock above the head(s) of the quarrymen
> was 100 cubits (Miller 1978, 217).

The Pool of Gibeon

The Bible mentions a "pool of Gibeon," where the forces of David and Saul met in battle (2 Samuel 2:13). In 1956, archaeologists in Gibeon uncovered a pit thirty-seven feet in diameter, cut from solid rock, which goes down eighty-two feet, all the way to the water table. Almost 3,000 tons of limestone would have had to be removed in the process. This pool apparently served as a water source for the village of Gibeon, since a seventy-nine-step stairway, also carved from the rock, winds along its side. From the rubble found at the bottom of the pit, archaeologists determined that the pit had not been used since 6 B.C. (Reader's Digest Editors *People* 1981, 168).

Water Tunnel at Megiddo

A similar shaft appears at Megiddo. There the problem resembled Jerusalem's: The water source, a spring, was outside the city. So in the time of Solomon, workers dug a shaft down 30 feet through the debris of previous settlements. From this rock, they built a stone wall, forming a well 15 feet in diameter and steps of stone leading to the bottom of the well. Then they continued digging through the rock, down to a level 120 feet below the spring. A 165-foot-long tunnel, mined from both directions, brought the spring's water downward to the base of the well (Paul 1973, 138).

Qumran

The Essene community at Qumran needed water not only for the necessities of daily living, but also for their baptism rituals. Thus they built their village around a complex canal system. An aqueduct brought water into the compound, where it branched off into conduits that supplied seven major cisterns, several bathing pools, and some buildings—and flowed off in a waterfall to the west (Reader's Digest Editors *People* 1981, 303).

Greek and Roman Waterworks

Herodotus described two of the great Greek engineering feats found on the island of Samos:

> One is a tunnel, under a hill 900 feet high, carried entirely through the base of a hill, with a mouth at either end. The length of the cutting is almost a mile— the height and width are eight feet. Along the whole course there is a second cutting, thirty feet deep, and three feet broad, whereby water is brought through

pipes, from an abundant source to the city. . . . Such is the first of their great works; the second is a mole in the sea, which goes all round the harbor, nearly 120 feet deep, and in length over 400 yards (Miller 1966, 168).

Explored by archaeologists in the 1870s and 1880s, the tunnel proved to be about two-thirds the size Herodotus had recorded—3,300 feet long and 5.5 feet in height and width. It had a trench for a clay pipe. Tunneling from both ends, the miners had apparently missed meeting at the middle by only twenty feet horizontally and three feet vertically.

The Romans, of course, claimed their water-supply engineering was far superior to anyone else's, and they were probably right. In the second century A.D., Trajan's water commissioner, Frontinus, trumpeted: ". . . With such an array of indispensable structures carrying so many waters, compare, if you will, the idle Pyramids or the useless, though famous, works of the Greeks" (De Camp 1963, 172).

The Romans began building their water system in 312 B.C., and the ancient world had never seen anything like it. By A.D. 226, eleven aqueducts were delivering 260 million gallons of fresh water each day to the city of Rome. More than 250 miles of aqueducts brought water from sources in the hills, along the hydraulic gradient, over canyons and valleys, into the city. Pressure from the elevated aqueducts forced water through a complex network of channels and conduits that ran under the streets of Rome (Babbitt 1939, 2, 3).

In that time, aqueducts became popular public works throughout the empire. It seemed as if every governor or provincial king wanted to put a modern Roman water system in his major cities.

In 9 B.C., Herod the Great installed an above-ground aqueduct for his favorite city of Caesarea, which had a population of 40,000. This aqueduct carried fresh water from the springs of Mount Carmel, over five miles away. The structure consisted of an elevated channel lined with plaster and supported by masonry arches. In some cases, concrete covered the channel, so for short distances the flowing water could be siphoned upward by hydraulic pressure.

It is obvious that, even here, the Roman engineers knew principles of hydraulic flow in open channels, since they generally built the aqueduct along the hydraulic grade line.

In Libya, a 100-mile aqueduct between Jebel and Leptis Magna carried not only water, but also olive oil. Poured in at Jebel, the oil rode atop the water until it was skimmed off at Leptis Magna and placed in separate containers for export.

Figure 127: A Stone Aqueduct
The ancient builders erected two parallel walls of brick or ashlar stone, which they lined with lime plaster. After filling in the lower portion with concrete or rubble, they laid a concrete floor and capped the aqueduct with flat stones (Landels 1978, 38).

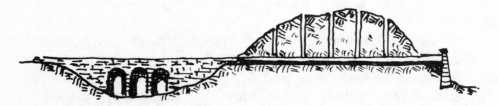

Figure 128: A Roman Aqueduct
In order to bring water where it was needed, the Romans built aqueducts that passed through land masses and over low places. Here one brings water from a higher, distant reservoir to a town that lacks a local water supply.

By building on a slope, the Romans made use of hydraulic pressure. The average angle in this case is 1 foot of height for every 200 feet of aqueduct.

At right a subterranean tunnel (*specus*) lies beneath vertical shafts (*putei*) that provided the miners with ventilation and allowed surveyors to keep construction on a straight course.

At left a stone arch (*arcuatio*) creates a level course for the aqueduct (Landels 1978, 40).

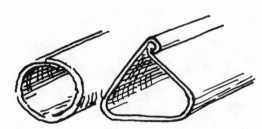

Figure 129: Roman Lead Pipes
During the empire, the Romans crafted pipes from lead sheets. After bending them to the proper shape, they lapped or folded the metal and soldered it closed with tin or an alloy of tin and lead (Landels 1978, 43).

Roman aqueduct construction required a great deal of tunneling. Early sewers and drains used a simple cut-and-cover method: dig a trench, install a masonry conduit, and backfill around it with dirt. But in some areas the water's flow line was too deep for cut and cover. Workmen would dig two shafts to the proper depth and then excavate tunnels across, connecting the two shafts by a single, deep tunnel. Tunnels were generally narrow, with room for only one workman. Here, too, they would dig simultaneously from two sides. Even with the use of plumb lines, however, the two tunnels did not always align.

Of course, workmen would often have to dig through rock. This could be grueling. Drilling and wedging, they would slowly break away the rock. Sometimes they alternately used fire and cold water to crack the rock's surface. Digging became even more difficult when workers encountered underground water. The only mechanism with which they could remove water was Archimedes' screw, and this would not work in the confines of a narrow shaft, so often the water would have to be lifted out in buckets (Armytage 1961, 19).

Sanitation and Sewerage

Healthy societies need not only a dependable water supply, but also a safe way of discharging waste water. Thus drains and sewers also became important building projects in ancient cities.

In Jerusalem, a drainage system carried off the water used in religious rites. Other sites had extensive storm sewers. While the royal latrine at Sargon's palace in Assyria (Dur Shariukin) had a water jar and clay dipper for flushing the facilities, the palace at Knossus, Crete, had a constant-flow flushing system. Latrine drains were usually separated from other systems so that noxious gases would not rise through system inlets.

Figure 130: Ancient Methods of Moving Water
To move small quantities of water, ancient man had machines that could
be worked by hand.

At lower right a man operates an Archimedes' screw. Inside a wooden
stave tube lies a solid spiral screw attached to a round timber shaft. The
upper and lower ends of the shaft are fixed to the ground and streambed
so that the entire apparatus can rotate. As the man turns the handle on top
of the shaft, the water is carried upward.

At top an Egyptian carries water from the river into an irrigation
channel by using a shadoof. Below him is a leather container, which he has
filled with water. A line attaches it to the beam above him. The pillar,
halfway down the beam, provides the fulcrum, and a counterweight or
centerpoise, at the end of the beam, balances the weight of the water
(Readers Digest Editors *People* 1981, 66).

In Roman times, city planners built sewers under the streets, rather than under buildings. In Caesarea, archaeologists found a drain opening in the center of a paved street. It appears to be a ten-foot-deep manhole, leading to an arched layer of brick—a remnant of an ancient sewer system.

In fact, Josephus commented on the elaborate construction of Caesarea's sewers. Half the time spent building this new city, he said, was spent on the sewer system. "Some of the sewers led at equal distances from one another to the harbor and the sea, while one diagonal passage connected all of them, so that rain water and refuse of the inhabitants were easily carried off" (Josephus *Jewish Antiquities* 15:340).

12
Travel and Transportation

From early times, the land of Palestine has been a crossroads. At the intersection of Europe, Asia, and Africa, it became the hub of ancient caravan routes between major civilizations.

The Via Maris began in Egypt, at Zoan, on the Nile delta, and followed the Mediterranean coast to Gaza and Megiddo. Then it continued north to Hazor and on into Syria.

The King's Highway lay just east of Palestine. Beginning at the Nile, at On, north of Memphis, it moved east to Elath, then north along the Transjordan plateaus toward Hazor and Damascus. The traveler could then continue northeast to Tadmor and east to Babylon and Ur, on the Euphrates. This is the route on which Moses led the Israelites before their entry into Canaan, after he promised the Moabites and Edomites that his people would stay "on the main road" (Numbers 20:17 TLB). In addition, a series of east-west roads connected Mediterranean cities to Nineveh, Assur, Dumah, and other cities in the current-day nations of Iran, Iraq, and Saudi Arabia.

Thus Israel's very geography made transportation a crucial part of its national life.

Animal Power

In biblical times, various animals transported goods and carried passengers— the wild ass (onager), donkey, ox, horse, and mule. The horse had greater speed, but oxen were twice as strong and cheaper to feed. Mules and donkeys tended to be less temperamental than horses and carried loads more willingly. The rocky terrain of central Palestine also cut down the possible speed of a journey, so mules and donkeys were generally preferred for large loads and long distances (Landels 1978, 13).

Throughout the Bible, horses suggest warfare and wealth. In prosperous times, Israel could afford to mount a cavalry. Escaping from slavery in Egypt, the Israelites went on foot, while the Egyptians chased with their horses and chariots. But in the golden years of Solomon, Israel's stables were full. "Solomon had forty thousand stalls of horses for his chariots, and twelve thousand horsemen" (1 Kings 4:26 KJV). In fact, Solomon conducted extensive trade, buying chariots from Egypt and horses from Asia, for export to Hittite and Aramean kings (1 Kings 10:29).

Wheeled Vehicles

The first "wheeled" vehicle was probably "invented" when someone edged a log under a heavy block of stone being moved from a quarry. As this procedure developed, workers would use rows of logs—when one slipped from the back, they'd place it in front. Eventually, wooden rails would be placed lengthwise in the path of the heavy object, and the logs would roll evenly upon that.

Later, the logs were cut into wheels, which would be attached to a wooden axle that would be connected to the cart holding the cargo. The wheel developed from a single slice of log to sections of a log pieced together. Still later, a band of metal set around the edge of the wheel made it more durable.

The Philistines mounted a cart on a single axle with two wooden wheels. The wheels consisted of sections of wood held together with a metal strip and surrounded by a metal rim. Babylonian ruins include copper wheel rims, but most civilizations, including the Philistines, used iron.

The carrying capacity of those ancient carts depended largely on the wheel design. The Philistines excelled in this area. They would cut three pieces from longitudinal sections of a log, split parallel to the grain. Then they fitted the two outer semicircular pieces around the inner piece, locking them together with metal straps and spikes. After heating up the metal rim, they placed it around the wheel. When it cooled, the rim shrank, locking the pieces of the wheel sturdily in place.

Though at first the Philistines used their carts merely for hauling, the neighboring Canaanites developed them for use in battle. Egyptians and Babylonians also began to build chariots. For success in battle, these carts needed greater speed and maneuverability, which the heavy, bulky wheels did not provide. So the Baby-

Figure 131: A Man-Drawn Cart
(Reader's Digest Editors *People* 1981, 122).

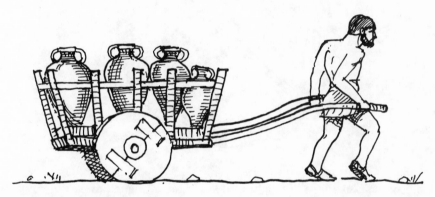

Figure 132: A Biblical Cart
The solid wheels were made of two or three sections of planking, cut from
the center of a log. The outer, soft sapwood, which would wear rapidly,
was removed before the planks were shaped, and wood battens held the
hard wood together (Reader's Digest Editors *People* 1981, 122).

lonians began taking sections of wood out of the wheel, resulting in spokes. The
Egyptians made rims of leather, which were connected with leather thongs. The
cart itself was streamlined into a mobile firing platform, built for speed and light
enough that it could be carried by two men over rough terrain. Some were
outfitted with quivers, bow cases, sheaths, and stands for axes and spears.

The chariot may have reached its peak with the Assyrian army of Ashur-

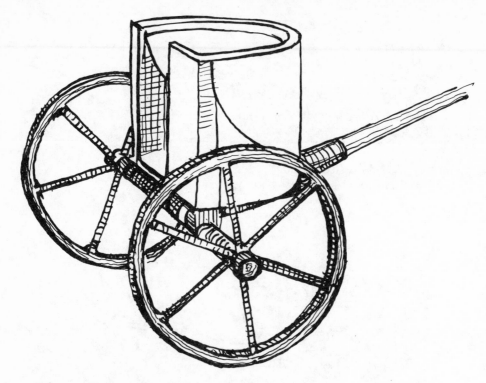

Figure 133: An Egyptian Chariot
The chassis was made of a light wooden frame, designed to carry two or
three warriors. Wheels were made of bent wood. Elm, oak, and ash
composed the hub, spokes, and felly, respectively.

banipal (885–860 B.C.). These large-wheeled vehicles carried four soldiers: a
driver, an archer, and two shield bearers.

The cart also was developed for nonmilitary uses. An ox-drawn wagon
could transport heavy loads of produce or equipment and thus was widely used
in agrarian cultures. Oxen usually pulled in pairs, attached to the cart by yoke
and pole. Since oxen have minor humps at the withers, a yoke fits them well.
Horses, shaped differently, required another arrangement. The flexible throat
harness used with oxen gave way to a stiff collar, which would not ride up on the
horse's neck, but put pressure on the shoulders and chest of the animal (Landels

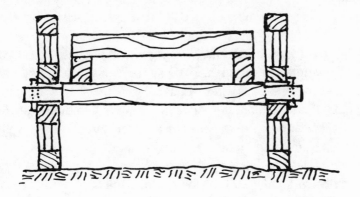

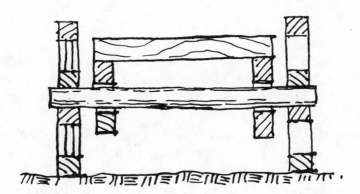

Figure 134: Section Through Two Carts, Showing Alternate Assemblies

The wheel at the top is held by a linchpin upon a fixed, stub axle, while the wheel at the bottom is fixed to a rotating axle. Parts of the wheel include the hub, spokes, and felly, with rims of steel or copper. The axle supports the cart and stake rails (Landels 1978, 182).

1978, 174; Reader's Digest Editors *People* 1981, 45). The harness connected to shafts that ran beside the animal and back to the cart, ensuring equal pull on both sides.

In general, ox-drawn carts carried cargo; horse-drawn carts carried passengers. It was strength versus speed. The same idea held true of wheel design. The heavier ox-drawn wagons often had solid wheels or a simple design of criss-crossed planks, while passenger carriages might use lighter spoked wheels. Carts could have either two or four wheels. Oxen fared rather well with two-wheeled carts, as long as the load remained balanced.

In lighter vehicles, wheels generally rotated on a fixed axle. A short stub of the axle would protrude beyond the wheel, and a linchpin would keep the wheel from falling off. To prevent excessive wear, a metal washer sometimes separated the wheel hub from the linchpin. In heavier vehicles, the wheels would be fixed on the axle, and the axle would rotate within brackets fastened to the bottom of the cart. (*See* figure 134.)

Roads

With the development of wheeled vehicles, the quality of roadways became a crucial concern. Single animals might poke along a primitive trail, avoiding ruts and obstacles, but a team of animals pulling a wagon needed smoother and more level paths.

For thousands of years, road building merely consisted of clearing away boulders and filling in the larger potholes. Most cities had narrow, muddy streets. A few had some paving of fieldstone slabs, but even these did not allow for drainage of rainwater. Some Greek roads followed the beds of dry rivers and streams. In areas where repeated travel had worn two deep wheel ruts into the dirt road, these ruts were paved with cut stone, switches, and siding. The only problem was that the widths of wagons and chariots varied—from four feet, six inches, to four feet, eleven inches, between the wheels—and some vehicles might not fit the grooves.

At first, paving was reserved for processional avenues. One sacred street in Nebuchadnezzar's Babylon had a base of large, flat bricks set in a mixture of lime, sand, and asphalt. On top of these lay a pavement of limestone flags. During religious processions, wagons would roll through this street, bearing statues of the gods. Paving of sacred ways was common in the ancient world, appearing as early as 1200 B.C. At Ashur, a processional road had grooves in the pavement for

Figure 135: Cutaway Section Through a Roman Highway
The sequence of construction includes excavation to a firm subgrade,
stone-rubble fill, flat slabs set in mortar, covered with crushed stone and
cement, curbs of cut stone, and flat paving stones closely fitted together,
forming the finished pavement.

the sacred wagons, assuring a smooth, safe ride for the gods (Armytage 1961, 18;
Miller 1966, 44, 45).

Josephus mentions that Solomon paved the roads leading to Jerusalem with
black stones. These, too, might be considered processional avenues. Yet in
Palestine the need for paving didn't arise until the tumultuous Persian and Hellenistic periods, when armies carted siege machines back and forth throughout
the land.

As with other areas of civil engineering, the Romans took road building to
its highest level. From about 400 B.C. to A.D. 200, the Romans built a vast system
of roads throughout their empire, covering a total distance of 75,000 miles.

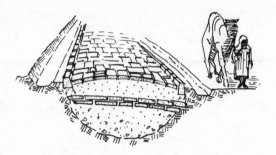

Figure 136: Section Through an Ancient Highway Between Towns

Figure 137: A Paved Road With Stone Curbs
(Cornfeld 1981, 250).

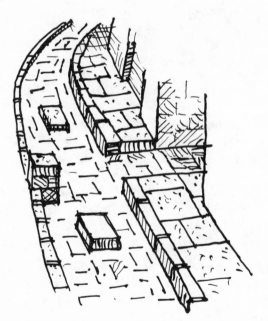

Figure 138: A Typical Town Road as Used Between Buildings
(Cornfeld 1981, 250).

Figure 139: A Roman Military Bridge
This bridge is constructed of braced trestle bents supporting wood beams

Figure 140: A Roman Pontoon Bridge
To provide troops with a way to cross to strategic positions, Roman
engineers spanned rivers with bridges built atop boats. Anchoring the
small vessels upstream, they made use of the force of the water to keep
the vessels heading in the same direction. They tied the boats together
with ropes and placed beams and planking on the gunwales to create the
floating roadway. During construction, they ferried workmen and materials
along the downstream side, on a raft.and planking. Bridge abutments were
made of wood cribbing.

Twenty-nine highways radiated out from Rome and connected with numerous other local roads. Many of these streets were paved with several layers of stone. Some of these stones can still be seen today.

Even in Palestine, archaeologists have found long sections of Roman-paved roads, complete with curbstones and milestones. Inscriptions on the mile markers list the name of the emperor, the official in charge of road construction, and the distance in Roman miles from the capital of the local district.

Roman roads were generally built and guarded by Roman soldiers. Rome had a sizable army and enjoyed long stretches of peace. At such times the soldiers became civil servants, carrying out a variety of public works, and the military engineers turned their attention to roads, bridges, and aqueducts (Bateman 1942, 1).

The Romans built their roads with a deep subbase of compacted rubble, covered with layers of flat stone slabs set in mortar. Then they laid down another level of concrete and crushed rock and on top of this the final paving of close-fitting stones. The surface of the road was slightly crowned, so water would drain off toward the curbstones (although some roads were inversely crowned, to drain toward the center). In most cases, drainage ditches on each side of the road carried off the rainfall.

The Romans had four distinct kinds of roads, from the one-foot-wide footpath (*semita*) to the eight-foot-wide highway (*via*). The Appian Way, begun in 312 B.C., was one of the first and most famous of the Roman *viae*.The ancient Roman road builders showed a great awareness of the crucial elements of road design: durable pavements, foundations, curbs, drainage, slope, and alignment. Their highways followed straight lines wherever possible, had deep cuts or fills to lessen excessive slopes, and passed over culverts and bridges when waterways had to be crossed.

Bridges

When King Darius I of Persia defeated Thrace and Macedonia in 512 B.C., he transported his troops over the Bosporus Strait by means of a floating military bridge designed by a Samian engineer, Mandrocles. A generation later, King Xerxes sent Persian troops over the Hellespont on a bridge that was supported by 100 ships.

These feats of engineering, forerunners of the modern military pontoon bridge, changed history. The time and effort saved with a well-placed bridge might give one army a significant tactical advantage.

The Romans knew this. They, too, built temporary military bridges, supported by boats or pontoons anchored in position against the current. For permanent installation, the Romans built arched stone bridges.

In sinking the piers for the stone-arch bridges, Roman builders would form a cofferdam around the place where they wanted it and then drive piles into the riverbed. Once the water was pumped out, the arch building could begin. When the pillar reached the waterline, the cofferdam was moved to the next pier site.

One disadvantage of this type of bridge was that the narrow arches restricted water flow and thus increased the velocity of the water; this in turn intensified the erosion process. Yet a number of Roman bridges have lasted to the present day. One stone-arch bridge still in use spans the Tagus River in Alcántra, Spain. Roman masons used sturdy blocks of granite and no mortar to build the seventeen-foot-high arches. They also put a fortified gateway in the middle of this bridge.

As we have seen, ancient engineering was based on intensive application of simple principles and an abundance of raw materials and cheap labor. The Romans devoted more energy and materials to civil projects than previous empires had. While others had constructed great palaces and tombs, the Romans devoted their efforts to roads, bridges, aqueducts, harbors, and sewers. Succeeding civilizations have benefited.

13
Ships and Harbors

W e have little information from ancient times about the precise designs and methods of shipbuilding. Perhaps the lore of shipbuilding stayed within certain guilds or families, which passed it on to each new generation by word of mouth. Still, from the many extant descriptions of ships and nautical journeys, we can trace the development of nautical engineering.

Egyptian Vessels

The Egyptians and Babylonians had riverboats long before ships sailed the Mediterranean or Red Sea. The earliest vessels, about 4000 B.C., were papyrus boats propelled by oars or a square sail.

When the Egyptians began crafting wooden boats, they were still lightly built, keelless vessels with only a few thin ribs. Planks were pinned to one another, rather than to a frame. A few beams ran from gunwale to gunwale, supporting a deck. This type of construction was fine for the peaceful Nile, but not for the Mediterranean, with its storms, harsh waves, and ripping currents.

One Nile boat did, however, manage to transport two obelisks, 100 feet long and weighing 350 tons apiece, from quarry to building site. The boat was so large that each of its steering-control oars weighed five tons. It took twenty-seven smaller boats, bearing thirty oarsmen each, to pull this vessel and its load.

Yet this feat shows that Egyptian engineers knew and used the laws of buoyancy and gravity. The ships's hull had to be large enough so that the weight of the water it displaced was greater than the weight of the ship and its cargo. The upward force of buoyancy would counteract the downward force of the obelisks' (and the ship's) weight.

About 1500 B.C., the Egyptians had merchant ships eighty or ninety feet long. Each wooden vessel had a taut hawser between the bow and stern, to reinforce the hull, since it still had no keel or ribs. When the Egyptians did build a fleet of troop-transport ships to sail the Mediterranean, they strengthened their design considerably. Shipbuilders would loop a large hawser around one end of the vessel, carry it across the ship's centerline, and then loop the other end. A stout pole was placed through the strands of the hawser, where it passed over the deck, and was twisted to tighten the rope like a tourniquet. The resulting tension compressed the ship. The hawser served the purpose of a keel and ribs, by holding the ship together amid the battering of the waves.

Builders also ran a horizontal netting around the upper part of the hull. This worked like a girdle, holding the ship together. Since a single mast required a sturdy keel and frame, the Egyptians used a two-legged mast, which distributed the weight across the hull. A single square sail rode on this double mast, stretched between two horizontal poles. Generally the ship also contained oarsmen, who could propel the vessel when the wind was calm.

Light ships like these proved graceful and fast, but they lacked the strength for extended sea travel.

Figure 141: Various Early Vessels

A large, oared riverboat (*right*) was used in Egypt about 2000 B.C.

Reed rafts or skiffs similar to the one at left are still used in the Lower Euphrates Valley. Workmen built them by tying reeds into bundles and lashing them together. By turning up the ends of the boat, they created a stern and bow. To make the craft more watertight, they could cover it with bitumen.

Egyptians used boats like the one at center when they ventured into the sea. The rigging served a double purpose, since it held the mast in place and allowed the crew to raise and lower the sail. Once the spar was aloft, they could furl the sail on the boom (Mansir 1981, 9).

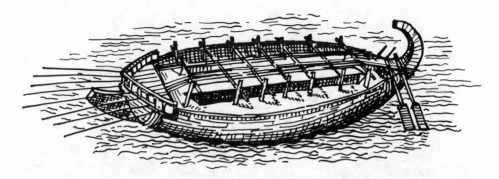

Figure 142: An Egyptian Riverboat
Unlike most modern boats, this ancient craft had no keel. Instead a huge
hawser and harnesses on the centerline of the ship held together its sawn
cedar planks with scarfs, which were tied together with rope fasteners.

With such a vessel the Egyptians carried huge stone obelisks along
the Nile.

Phoenician Ships for Commerce and Warfare

Egyptian seapower waned about 1100 B.C. The Phoenicians took their place,
dominating Mediterranean trade for centuries.

The Phoenician city-state of Byblos (now Jubayl) became one of the Mediterranean's great shipping centers. "Byblos ships," as they were called, carried
cedar wood from Lebanon to Egypt and came back with papyrus and agricultural
produce to ship elsewhere. (In fact, the Greeks used to call the scrolls of papyrus
that came from this "biblos"; from that we get our words *Bible* and *bibliogra-
phy*.) Phoenician ships sailed as far as North Africa, the west coast of Spain, even
southern England. Their seafaring strength lasted until the time of Alexander
(Miller 1978, 362–365).

The Phoenicians used two kinds of ships. The "round" ship, with its rounded
hull and a raised prow and stern, was used exclusively for commerce. The larger
and sleeker "long" ship measured more than 100 feet in length and could be used
for either commerce or war. It had very high decks and curved prows, two banks
of oars (sixty in all), two steering oars, and a mast in the center that bore a large,
square sail. In war, it might be outfitted with bronze shields and a pointed ram
and its bow covered with bronze or iron. The long ship could hold as many as

Figure 143: A Wooden Seagoing Ship
This typical ancient Phoenician merchant was outfitted with both sails and oars. Generally these ships sailed northward, with the wind, and used oars for going southward, against the wind.

250 men. The Phoenicians were the first to send a convoy of warships with their merchant fleet.

The prophet Ezekiel described the glory of the Phoenician seaport of Tyre:

> O mighty seaport city, merchant center of the world, the Lord God speaks. You say, "I am the most beautiful city in all the world." You have extended your boundaries out into the sea; your architects have made you glorious. You are like a ship built of the finest fir from Senir. They took a cedar from Lebanon to make a mast for you. They made your oars from oaks of Bashan. The walls of your cabin are of cypress from the southern coast of Cyprus. Your sails are made of Egypt's finest linens; you stand beneath awnings bright with purple and scarlet dyes from eastern Cyprus.
>
> Your sailors come from Sidon and Arvad; your helmsmen are skilled men from Zemer. Wise old craftsmen from Gebal do the calking. Ships come from every land with all their goods to barter for your trade.
>
> Ezekiel 27:3–9 TLB

The united kingdom of Israel reached its economic and territorial peak in the tenth century B.C., under King Solomon. Solomon shrewdly allied with the Phoenician ruler, King Hiram of Tyre. Using Israelite seaports, Phoenician know-

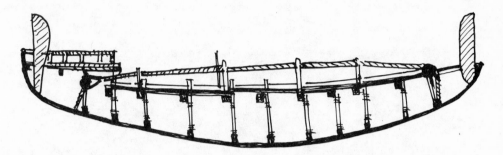

Figure 144: Section Through an Egyptian Ship
Many keelless ships, like this seventy-foot one, used only a hawser for
support. Cargo would be stored on the deck. A single or double mast
would be mounted midships. At the left, in the stern, lies a raised platform,
which acted as the bridge or poop.

how, Israelite wealth, and Phoenician timber, Solomon built a navy that conducted lucrative trade throughout the known world.

> King Solomon also built ships at Ezion Geber, which is near Elath in
> Edom, on the shore of the Red Sea. And Hiram sent his men—sailors who
> knew the sea—to serve in the fleet with Solomon's men. They sailed to Ophir
> and brought back 420 talents of gold, which they delivered to King Solomon.
>
> 1 Kings 9:26–28 NIV

> The king had a fleet of trading ships at sea along with the ships of Hiram.
> Once every three years it returned carrying gold, silver and ivory, and apes
> and baboons.
>
> 1 Kings 10:22 NIV

The ships of this Israeli–Phoenician fleet probably had high prows and
single masts that bore single, rectangular sails; they may also have had one bank
of oars on each side. Obviously these vessels could sail great distances.

However, the places they went to are something of a mystery to modern
scholars. No one knows for sure where Ophir was, though some have speculated
it was East Africa or even Sumatra, and Josephus said it was in India. Tarshish
appears often in the Bible, and is usually described as a source of metals such
as tin and copper. The word may have meant "metal refinery" in ancient Phoe-

Figure 145: A Phoenician Round Ship
Also known as the Hippo ship or Tarshish ship, this vessel had a square
sail reinforced by bunt lines attached to the foot of the sail and by loops
woven into the fabric.

nician and thus might refer to any port with Phoenician smelters. Some scholars
have placed Tarshish in Spain (Reader's Digest Editors *People* 1981, 184).

Greek Sailing Innovations

Homer described the vessels of the Trojan War as "hollow ships." These must
have been mostly undecked, except for a small landing that could hold a lookout
and a few marines, when the ship went to war.

With both sail and oars, such ships could make the best of any fine weather.

If a favorable wind blew, the crew erected the sail, first setting up the mast with two forestays, stepping it, locking it with a wedge, and securing it with a back-stay. Once they hoisted the sail, they set it with braces. Then the helmsman would take his position, holding the leeward sheet in one hand and the tiller in the other. Some ancient vessels might also carry a foresail (artemon), beneath a slanting bowsprit, to keep the bow downwind.

If the wind failed, the crew pulled oars from under the benches that ran along the gunwales. With leather straps, they attached these to the tholes (wood-en pins used as oarlocks, which provided a fulcrum against which the oar could work).

But this did not solve every sailing problem; most ancient sailing needed the wind—oars could only take you so far. So ships often sat in port, waiting for the right wind to blow. In bad weather, especially from mid-November to mid-February, when winter storms ruled the Mediterranean, sailing also became dangerous, as shown by Paul's experiences in Acts 27, and a good sailor might rightly fear going to sea.

By 800 B.C., Greek shipbuilders created the fighting vessel that would be

Figure 146: An Ancient Hollow Ship

state of the art for the next thousand years. Homer's old sea rovers, useful in trading and piracy, were inefficient in battle. The new warship had a high, rounded stern and a straight prow with a strong, pointed ram. This ram changed the face of warfare at sea. No longer were ships just transporters of archers and spearmen—now the ship itself was a weapon. The navy with the skill to maneuver ships into attacking positions would emerge victorious. Battle became a matter of precise timing. A slight miscalculation might be the difference between ramming the enemy and having your own hull ripped open by the enemy's ram.

But warfare wasn't left entirely to the ship. The new vessel also included a fighting platform for archers and spearmen. A deck ran from stem to stern, but not all the way across from bulwark to bulwark. Space was left at each side for rowers to work at or below deck level. While the ship cruised, the oarsmen would sit at deck level, but in battle they would retreat to a more protected position below deck. The opening in the deck provided ventilation.

As ships grew larger, they needed more power. The answer lay in the addition of a second bank of oars on each side. The Phoenicians originally invented the two-bank galley ship, or bireme, but the Greeks developed it further and used it widely by 500 B.C. (Mansir 1981, 17).

Later, the Greeks and Romans installed a third bank of oars. This required additional manpower, but despite popular conception, the oarsmen were not all slaves. Most were freemen. These amazingly maneuverable triremes could reach a speed of ten knots and spin in tight circles. Yet they were not especially seaworthy. With only an eight-foot freeboard and a three-foot draft, a trireme would be in trouble in turbulent seas. Sails still carried these ships on the high seas, but in battle the rowers supplied the power. Some triremes were also equipped with an underwater metal-tipped ram.

Roman Ships

Romans used two major types of sailing crafts: long ships and wide ships. The long ships—triremes and quinqueremes (which had five banks of oars)—were narrow and swift, used mainly for war. The heavy wide ships carried cargo and passengers.

The Roman navy ruled the Mediterranean, and with Rome's warships patrolling, keeping the peace, sea trade flourished. Merchant vessels from Greece, Phoenicia, and Egypt crossed the Mediterranean unmolested.

Figure 147: A Greek Pentecenter
This ship consisted of the prow at the bow; a bridge poop, aft; and a mast
at midships, supporting lapped boom yards and stayed by fore and aft
braces (Landels 1981, 153).

The Apostle Paul traveled on trading ships, since in his day there were no
vessels built specifically for passenger service. Merchant ships would routinely
take on several dozen passengers.

A typical merchant ship might be 140 feet long, 36 feet broad, and 33 feet
deep—with a capacity of three tons. Its design would enable it to sail close to the
wind. The decks were covered, with cabins below. Built of wood, with a simple
prow, it might have a raised aft that inclined toward the center of the ship.
Though some used oars, most trade vessels relied on sail power. Early ships had
a single square sail on a fixed mast, but during the Roman Empire two or three
triangular sails propelled the vessels (Reader's Digest Editors *People* 1981, 390).

Figure 148: A Bireme or Two-Banked Galley
At the stern, one of a pair of steering oars guides the vessel. Forestays
secure the mast to the bow, while brails hold it to the stern. Ropes
anchored to both sides of the bow keep the yards in position.

On orders from Emperor Caligula, the Romans once hauled a huge stone
obelisk from Egypt on a specially designed ship that held 1,300 tons. Taken from
Hieropolis, near Cairo, the monument was barged down the Nile and shipped
across the Mediterranean on this special vessel. Then it was barged up the Tiber
to Rome, where it was reerected.

Navigation

Once people began to sail beyond the sight of the shore, they needed a way to
find out where they were. The earliest navigators may have judged their location
from the wind direction or by watching the waves or even the birds.

Eventually, sailors found Polaris, the north star, also called the pole star.
This star did not appear to move. It stayed there, always in the north, a true guide
for navigators.

Figure 149: Section Through a Trireme or Three-Banked, Galley
This fifth-century Athenian vessel shows the principal elements of
fighting-ship construction used during the Greek Golden Age. Along the
spine of the ship, the keelson provides support from bow to stern. A rib
frame surrounded by a plank hull holds the three levels of oarsmen, and a
raised platform covers them. Atop the mast, a square sail of sheet linen
would be suspended with brails on the lapped yards attached to the mast.

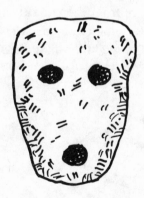

Figure 150: A Large Stone Anchor
This shaped stone anchor from about 1200 B.C. was
found in Caesarea.

The Phoenicians learned to measure their position with their hands. Extending an arm, the ancient navigator would line up his little finger with the horizon and his thumb with the zenith. With certain calculations, he could judge his direction and latitude.

Possibly about 150 B.C., Hipparchus invented the astrolabe. Literally a "star taker," the astrolabe measured the angle of the stars above the horizon. With this a sailor could determine nautical position. Hipparchus's invention consisted of a movable pointer fixed to the center of a wood or metal disk, which had its circumference marked off in degrees. This was the predominant navigational instrument until the sextant was invented.

Navigators might also probe the sea bottom with a sounding pole or a tallow-tipped lead and line, to determine the depth and nature of the sea floor at that point. For instance, Herodotus wrote in the fifth century B.C., "When you get eleven fathoms and ooze on the lead, you are a day's journey from Alexandria" (Casson 1959, 68).

Two thousand years before Columbus sailed to America, Pythagoras altered man's understanding of navigation when he proclaimed that the earth was round. Later, both Plato and Aristotle would reassert this theory. In the third century B.C., Eratosthenes determined the circumference of the earth to be 25,000 miles, by using fairly simple geometric calculations. In the second century A.D., Claudius Ptolemy laid out the first terrestrial coordinates, the parallels of latitude and the meridians of longitude. The parallels measure the distance from the equator in degrees, up to 90 degrees at the poles. The meridians determine one's east-west position, measured in degrees (up to 360) from a fixed "prime" meridian. With Ptolemy's coordinates, navigators could more precisely plot their position at sea.

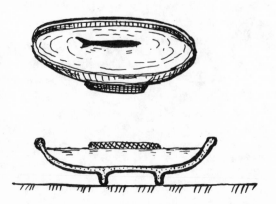

Figure 151: An Ancient Magnetic Compass

Early magnetic compasses like this one, used in the Far East, consisted of a clay bowl containing water upon which floated a thin leaf of magnetized iron (or lodestone) shaped like a fish. The head of the fish would always point to magnetic north (Institute of History 1983, 154).

Harbors

Because early ships could not travel long distances at one stretch, numerous ports were established around the Mediterranean. As sea trade flourished, many ports became significant economic centers.

These cities needed efficient harbor facilities to welcome ships and their commerce. Building harbors was, in many cases, as important as the construction of fortresses. Harbor design required not only solid engineering knowledge, but also an awareness of the tides and the destructive effect of the waves (Morgan 1960, 162).

One of the most ancient ports on Palestine's coast, Joppa (or Jaffa) bears the remains of numerous civilizations. Excavators have found fortifications from the Bronze Age right on through to Byzantine and Arab times.

Into Joppa's port came the cedars of Lebanon that Solomon used for the Temple construction (2 Chronicles 2:16). King Hiram's workers floated them down to the sea on barges. At Joppa Jonah boarded the ship for Tarshish, running away from Nineveh, where God had ordered him to go.

Acco (also called Acre or Ptolemais) was another major port of Palestine. North of Caesarea, but south of Tyre, it served the Jezreel Valley and the region of Galilee. Through Acco passed the major highway from Egypt, the Via Maris, and east-west roads through Jezreel connected Acco with the other major thoroughfare, the King's Highway.

Gaza, in the south, originally served as a key caravan point between Egypt and the rest of the Middle East. After the Romans rebuilt the city closer to the shore, this harbor became a major port of embarkation for eastern merchants.

Tyre was once an island, one-half mile from shore. In 332 B.C., the army engineers of Alexander the Great built a land bridge or causeway so they could bring their siege engines to the city walls. Over time, silt has accumulated on this land bridge to form the present neck of land, 600 yards across at its narrowest point.

Both Tyre and its sister city, Sidon, have small islands and reefs offshore that function as breakwaters, creating a peaceful inner harbor.

Caesarea, however, emerges as the most magnificent port built on Palestine's coast. Originally known as Strato's Tower, named for a Phoenician king, Herod renamed it to honor the emperor. He spared no expense in making this a world-class seaport.

Since Caesarea had no natural harbor, Herod's engineers built a 200-foot-wide breakwater for docking and anchoring vessels; it also protected the inner harbor from the ravages of waves and currents from the southwest. To the north, a semicircular seawall enclosed a marina of sorts—a yacht basin with an entrance guarded by two immense, statue-bearing towers.

The northern breakwater measured an estimated 250 yards, the southern one 600 yards. The enclosed area measured about forty acres. In his *War of the Jews*, Josephus described the construction of Caesarea:

> After Herod had marked out the comparative dimensions of the harbor
> . . . he had blocks of stone sunk into twenty fathoms of sea—most of which
> were fifty-feet long, nine deep and ten broad, some being even larger. When
> these submarine foundations had risen to the water level he built above the
> surface a mole 200 feet wide; one hundred of these are built out into the sea
> to break the force of the waves and, therefore, called the breakwater, while
> the remainder supported the stone wall that encircled the harbor. Massive
> towers are spaced at intervals above this wall and the most conspicuous and
> magnificent is called Drusium after Caesar's stepson, Drusas (Whiston 1979,
> 162).

Apparently, the builders began by digging out the inner harbor and piers. Once they had excavated the outer harbor, too, they removed the narrow strip of land separating the two harbors, allowing water into the inner basins.

At the bottom of the harbor, we find a mastery of harbor-building technology that rivals present-day knowledge. At several points a sluice system enabled sand-free water from the open sea into the enclosed harbor. This created circulation within the enclosed basins and inhibited silting. A subsidiary breakwater system protected the main enclosing arm of the harbor at its most vulnerable point.

These builders also made great use of the new hydraulic concrete that hardened underwater, becoming practically impervious to the sea's destructive force. At Caesarea archaeologists have found blocks of hydraulic concrete weighing fifty tons. They measure fifty feet by nine feet by ten feet, even larger than those Josephus described.

Completed in 9 B.C., the city had a population of 40,000 by the second century A.D. Caesarea was the first port in Palestine that could provide anchorage for as many as one hundred Roman war galleys. The moles contained buildings that served as dormitories for sailors and warehouses and had walkways where people could enjoy the sea breeze. A tall tower served as a lighthouse.

In A.D. 130, an earthquake damaged the harbor, and by the fourth or fifth century A.D. it had become unusable.

Trade and the Biblical World

Not until the reigns of Solomon and Jehoshaphat did the Israelites develop extensive land and sea trade routes. These two kings, who ruled almost a hundred years apart, utilized Israel's central geographic position as the land bridge between ports on the Mediterranean and Ezion geber on the Red Sea. But after neighboring powers conquered them, the Hebrews lost access to the ports and relied on land travel by caravan.

The conquest of Jaffa by the Hasmonaeans again provided the Judean kingdom with access to the sea routes when this port was opened to Greek merchants, who replaced the Phoenicians. Jaffa's port again became an important center of international commerce.

During the Roman period, trans-Mediterranean trade was controlled by the Roman fleet. Their large and strong merchant ships carried soldiers, goods, merchants, and passengers; some of the apostles traveled to and from Judea.

Judea become part of the Roman commercial complex. Nabataean trade, carried by camel trains, crossed southern Palestine, carrying spices, perfumes, herbs, precious woods, and gems from the Far East and southern Arabia to Gaza and other ports on the Mediterranean. By land and sea, the Holy Land continued to serve as the hub from which culture, knowledge, and religion spread throughout the region.

Glossary

Acanthus leaves: Architectural ornaments used in capitals of Corinthian and composite orders, representing the foliage of the acanthus plant.

Acco (also known as Acre and Ptolemais): For centuries the best and most important harbor in northern Palestine, it was a major stop on the main north-south highway, the Via Maris.

Adobe: Natural, sun-dried clays with a binder.

Adz: An ax-type tool used in chipping the surface of wood. Its cutting edge is perpendicular to the axis of the handle.

Altar: A place of sacrifice, built of a heap of stones and earth. Some altars were even made of acacia wood and overlaid with bronze.

Amorite: People descended from the fourth son of Canaan. They made up the pre-Canaanite population of Palestine.

Amphiprostyle: A building with columnar porches at both ends but without a peristyle.

Antonia Fortress: A structure at the north side of the Temple Mount, built by Herod the Great.

Aqueduct: A raised masonry structure containing a channel for carrying water from a source at a higher elevation.

Arcade: A series of arches supported by piers and columns, or a covered passageway.

Arch: A structure composed of separate wedge-shaped stones or bricks arranged in a curved line, so as to retain their position by mutual pressure.

Ark: A decorated chest in which the Tablets of the Covenant were kept. *Noah's ark:* a vessel built by Noah for saving his family's lives during the Great Flood.

Architrave: The main long horizontal beam supported by the upper slab of the column capitals, which replaced the wood beams.

Artemon: A small sail attached to the long spar forward of the bow, used to keep a ship before the wind and as a steering aid.

Ashlar masonry: Masonry constructed of square-hewn stone.

Assault tower: A wooden structure used to provide access to the top of a fortress wall or to support a battering ram for breaking through the wall.

Astrolabe: An ancient instrument, said to be invented by Hipparchus in 150 B.C., used for taking the altitude of the sun or stars at sea. The quadrant and sextant replaced it when navigational technology advanced.

Awl: A pointed tool used to chop wood. It has a wooden handle and a bronze or iron blade. A small boring tool.

Ballista: A siege engine that propels stones or iron darts.

Barrel vault: A masonry, archlike structure.

Battering ram: An ancient offensive engine of war used to beat down walls of besieged places. It consists of a large pole, with a head of iron shaped to resemble the head of a ram.

Battlement: A fortress, or a notched or indented parapet of a tower or wall used for defense.

Bireme: An ancient Greek or Roman galley with two banks of oars, originally developed by the Phoenicians.

Bitumen: Asphalt or tar.

Block: A pulley used with ropes and tackle, consisting of a slim oval shell of wood. Inside, ropes run around one or more sheaves (pulley wheels).

Bloom: Semiplastic molten iron.

Brass: An alloy of copper and zinc. In biblical usage the word *brass* signified copper or bronze.

Breakwater: A structure protecting a shore area, harbor, anchorage, or basin from the waves.

Bronze: A metal made from an alloy of copper and tin.

Bronze Age: A historical period characterized by the use of bronze for tools and weapons.

Buttress: A structure built against or projecting from a wall or building for the purpose of giving it stability.

Byblos: A port of ancient Lebanon.

Byzantine: Pertaining to Byzantium, an ancient city, later called Constantinople, which became the capital of the Eastern Roman Empire.

Caesar: The title used for the Roman emperors from Augustus to Hadrian, later applied to the heir presumptive.

Caesarea: Originally known as Strato's Tower, Herod the Great built it into a thriving port. The port was known for its extensive breakwaters and harbor facilities.

Capital: The head or cornice of a column.

Canaan: Part of ancient Palestine, between Jordan, the Dead Sea, and the Mediterranean; the land God promised to Abraham.

Capstan: Mechanism used for hoisting blocks or cargo, consisting of a vertical axle, drum, toothed wheel, and pawls fixed to the axle, which engage the toothed wheel to prevent a cable from running backward in the absence of pressure on the levers.

Casemate wall: Two parallel walls, a heavy outer wall and a lighter inner wall, connected by cross walls, which form chambers, or casemates, between the two walls. As an added defense, these chambers were sometimes filled with stones.

Catapult: A siege engine for throwing pikes or stones. A large crossbow mounted on a pedestal shot darts up to six feet in length. Also known as quick firers.

Causeway: A street or highway, a raised road across wet or marshy ground or across water.

Centering: Shoring for construction of arches.

Chariot: A two- or four-wheeled cart pulled by horses and used in warfare.

Cistern: A man-made reservoir that is filled with rainwater or water brought from springs.

Citadel: A fortress or castle.

Cherub: An angelic being illustrated in ancient art with a woman's head on a winged lion's body.

Colonnade: A series of columns placed at regular intervals and used to support a roof structure.

Conduit: A pipe, tube, or channel used for conveyance of water.

Corbeling: An overlapping arrangement of stones, in which each course projects beyond the one below.

Corinthian order: A style of Greek or Roman architecture decorated with acanthus leaves.

Cornice: The uppermost member of an entablature, resting on a frieze.

Counterpoise: A counter-balancing weight.

Coarse rubble: Masonry work made of undressed stones of irregular size.

Coursed masonry: Masonry work laid in horizontal planes.

Crane: A lifting apparatus for raising blocks or cargo.

Crenelated walls: A fortress wall with openings through which defenders could shoot on those below.

Crete: Also known as *Cheretim,* the largest and southernmost of the Greek islands, from which the Philistines may have originated.

Crossbow: A weapon formed by placing a bow perpendicularly on the stock, from which arrows are released.

Cubit: A biblical unit of measurement, 1 foot, 6 inches, or 1 foot, $9\frac{3}{8}$ inches; the length of the lower arm from the elbow to the tip of the middle finger.

Cut stones: Stones with smoothly dressed beds and joints.

Cyclopean masonry: Rubble masonry made with very large, irregular rock.

Dart thrower: A type of catapult.

Debir: The Holy of Holies of the Jerusalem Temple.

Dentils: A series of small, rectangular blocks, protruding like rows of teeth, as under a cornice.

Derrick: An apparatus used for hoisting and lowering weights, comprised of a boom secured at the foot of the mast, from which runs a block and tackle, controlling the elevation and traverse of the boom.

Displacement: The weight of a vessel based on Archimedes' principle that it equals the weight of water displaced by its hull.

Divided Kingdom: The period of Hebrew history from 928 B.C. to 586 B.C., ending with the destruction of the state of Judah and the Babylonian Exile.

Dome: A hemispheric roof constructed of shaped stones.

Dome of the Rock: An Islamic structure built on the Temple Mount.

Doric order: A style of Greek or Roman architecture.

Draft: The depth of a vessel from the keel to the waterline.

Eggs-and-darts: An architectural ornamentation.

Elath (Eloth):   A port on the Gulf of Elath.

Engines of war: Ancient artillery, machines used to hurl arrows, spears, and stones, including catapults and testudos.

Extrados: The exterior curve or surface of an arch or vault.

Ezion geber: A port at the southern point of Palestine, where Solomon's ships, built with Tyrian help, took part in independent maritime commerce.

Fathom: A unit of measurement of water depth, equal to six feet, or 1.83 meters.

Fiber rope: Ropes made from pine-tree bark, coconut hair, camel hair, horsehair, thongs cut from hide, cotton, jute, sisal, flax, or hemp.

Finger: A unit of measurement, .7 or .9 inch.

Forecastle: The forward deck, which is raised above the main deck. It was used as a fighting "castle" by soldiers carried on board the ships.

Frieze: The part of an entablature of a column that is between the architrave and the cornice.

Frontinus: A Roman engineer (A.D. 40–103) who built aqueducts in Rome.

Fulcrum: The point of support by which a lever is sustained.

Galley: An ancient sail- and oar-propelled vessel.

Gangue: Impurities in iron making.

Gastraphete: The belly weapon, a type of crossbow.

Gate post: The post upon which a gate swings or the one against which it closes.

Gaza: Town on the southwest coast of Palestine. It was an important caravan point en route to Egypt.

Gneiss: A metamorphic rock similar to granite.

Glacis: A sloping bank or defensive wall used against battering rams and to make approach more difficult. It also effectively ricocheted stones dropped by the defenders.

Granary: A structure or pit used to store grain.

Grapnel: A hooked device used to fasten and hold one ship to another.

Greek fire: A flammable mixture of naphtha, sulfur, and pitch used in warfare.

Groined vault: A structure made by the intersection of two barrel vaults.

Gunwale: The upper edge of a ship's side.

Gypsum: A mineral ($CaSO_4 \cdot 2H_2O$) used to make plaster.

Hand: A unit of measurement equal to four fingers.

Hasmonaeans: The family name of the priestly family popularly known as the Maccabees.

Hittites: An ancient people who flourished in the eastern Mediterranean area for about seven centuries before 1200 B.C.

Holy of Holies: The inner chamber of the sanctuary of the Jewish Temple.

Igneous rock: Original rock including granite, basalt, and diorite.

In antis: With a porch having two or more columns enclosed between projecting side walls of a cella (a windowless structure).

Intrados: The interior curve of an arch.

Ionic order: A style of Greek or Roman architecture.

Iron Age: A period in which use of iron superseded bronze.

Joppa: A port, also known as Jaffa, where Solomon brought timbers shipped from Lebanon on barges and had them hauled up to Jerusalem (2 Chronicles 2:16).

Josephus: A Jewish historian, public official, and general (A.D. 37–100). He was also called Flavius Josephus.

Keel: The backbone of a ship, from which the ribs project.

Kiln: A type of oven used to bake or dry clay or brick.

Knot: A unit of speed used in navigation, equal to one nautical mile (6,076.115 feet or 1,852 meters) per hour.

Lanyard: The rope trigger of a siege engine, catapult, or artillery.

Lime: Calcium oxide (CaO), a white, caustic solid made by calcining limestone and other calcium carbonate substances. It is used for mortar and cement.

Limestone: A rock consisting chiefly of calcium carbonate.

Lime plaster: Plaster made of dehydrated, crushed limestone.

Lintel: The beam over a door or window opening.

Long ship: A large, streamlined ancient Phoenician ship. (*See also* Trireme and Quinquereme.)

Mason: One who hews stones into shape for construction.

Menorah: A seven-branched candelabrum.

Meridian: A longitude line, a circle drawn around the earth, through the poles.

Milestone: A cylindrical, inscribed stone placed along the highway by the Romans. It describes the emperor authorizing the construction of the road, the person in charge of construction, and the distance in Roman miles to the capital of the district.

Military engineer: A soldier whose specialty is building and destroying military structures.

Mole: In coastal terminology, a massive land-connected, solid-fill structure of earth, (generally revetted) masonry, or large stone that serves as a breakwater or pier.

Mortar: A mixture of lime, sand, ash, and water, used for plastering cisterns and reservoirs to make them water resistant. It is also used as a binder in masonry.

Mount of Olives: A hill on the east side of the Temple Mount.

Obelisk: An ancient Egyptian stone monument.

Onager: A large siege machine, based on torsion, with a lever that pulled back against a tightly twisted rope of fiber, sinew, or hair. It was named after the wild ass because of its kick.

Parallels: Lines of latitude. The distance north and south of the equator, measured in degrees, from 0 at the equator to 90 at the poles.

Pentecenter: An ancient Greek sailing ship.

Peristyle: A structure surrounded by columns.

Philistia: An ancient country on the southwest coast of Palestine.

Phoenicia: Ancient Lebanon.

Pila: A thrusting spear.

Pillar: A vertical stone structure used to support or a column.

Pitch: A viscous substance used for waterproofing. It is made from coal, tar, asphalt, or wood resin.

Plaster: Mortar or cement used as a coating over clay or stone wall or as a binder between bricks and blocks. In Palestine, the material commonly used was clay, in the Euphrates Valley, slime (or bitumen).

Polaris: The north or polar star, by which navigators could figure their position.

Portico: A kind of porch fronted with columns, often at the entrance of a building.

Post and lintel: A type of structure involving two columns and a beam (or lintel).

Pozzolana: A volcanic ash or powdered rock used by the Romans to make hydraulic cement.

Prostyle: Having columnar porches at both ends, but no peristyle.

Proto-aeolic: The original or first.

Ptolemaic Period: The era of the Greek philosopher Ptolemy or the Ptolemies, kings of ancient Egypt.

Pulley: One of the simple machines for raising weights, consisting of a small wheel that moves around an axle and has a groove cut in its circumference, through which a cord runs.

Quay: A harbor structure, stretch of paved bank, or solid, artificial landing place parallel to a navigable waterway, for use in loading and unloading ships.

Quick firer: A type of catapult. (*See also* Catapult.)

Quinquereme: A galley with five banks of oars.

Ram: A siege machine used for battering down fortress walls. (*See also* Battering ram.)

Ram tortoise: A wheeled shed, or tortoise, with a battering ram suspended on chains hung from the roof.

Rigging: Ropes and other equipment used for hoisting and bracing of poles.

Rise: The vertical height of an arch, between the spring line and the crown.

Rubble masonry: A wall made of uncut stones.

Sand lime: A mixture of sand and lime, used as mortar.

Sapper: A soldier in the engineering corp, who is trained in fortifications.

Saw: Ancient saws were long flint knives with jagged edges or bits of flint fitted into a wooden frame.

Scorpion: A type of catapult.

Semicircular arch: An arch of less than 180 degrees.

Semita: A Roman footpath.

Sextant: An instrument for measuring angles of heavenly bodies from the horizon. Used in navigation.

Sidon: A port town on the coast of Lebanon.

Siege tower: A tall structure used to attack a city or fortress. The height gave archers a chance to fire down upon defenders. Some siege towers also carried battering rams.

Skein: Twisted fibers, hair or rope, forming an energy-producing device.

Skewback: A sloping surface that supports the end of an arch.

Sling: An offensive weapon used by hunters, by shepherds against wild beasts, and in warfare.

Spring line: The horizontal line at the base of an arch.

Stele: An upright slab or pillar, usually with an inscription.

Strato's Tower: *See* Caesarea.

Sumer: The ancient name of Babylonia.

Synagogue: A Jewish place of worship and house of assembly.

Tabernacle: The tent in which Hebrews worshiped in the presence of God and which was taken with them on their journeys in the wilderness.

Talent: A weight equivalent to forty-seven or forty-eight pounds.

Tar: A distillation of wood or coal employed to waterproof wood, textiles, yarns, or rope.

Tarshish: A biblical city from which copper was obtained.

Testudos: "Turtles," catapults that hurled flaming spears.

Tin: A metal used, along with copper, to make bronze.

Torsion machine: A siege weapon such as an onager.

Tortoise: A type of siege engine.

Trebuchet: A type of catapult.

Trireme: A battleship containing three banks or tiers of oars, used by the Greeks and Romans.

Tyre: A port city on coast of Lebanon.

Ulam: The outer chamber of the Jewish Temple.

United Kingdom: The period from last days of David, 973 B.C. to 933 B.C., when the Israelite kingdom was divided.

Unsquared Stone: Stone that is used as it comes from the quarry, without any preparation other than removal of very acute angles and excessive projections.

Vault: An arched structure, made of stones, forming a ceiling or roof.

Via: A Roman highway.

Via Maris: A Roman highway in ancient Palestine.

Vitruvius: Roman engineer-architect of the first century A.D.

Volute: A spiral, scroll-like, or twisted formation, a distinctive feature of the Ionic column.

Voussoirs: Any of the pieces in the shape of a truncated wedge that form an arch or vault.

Watchtower: A fortified structure on the wall or above the gate of a city.

Windlass: A hoisting or hauling apparatus consisting of a horizontal drum on which is wound a rope, attached to the object to be raised or moved. Used to weigh anchors on ancient ships, it was smaller than a capstan and operated on a vertical axis.

Ziggurat: A stepped Babylonian temple.

Bibliography

Allsopp, Bruce. *A General History of Architecture, From Earliest Civilization to the Present Day.* London: Pittman & Sons, 1965.

Angelucci, Enzo and Attilo Cacuri. *Ships: A History With Over 1000 Illustrations.* New York: Greenwich House, 1977.

Armytage, W. H. G. *A Social History of Engineering.* London: Faber and Faber, 1961.

Babbitt, Harold E. and James Dolland. *Water Supply Engineering.* New York: McGraw-Hill Book Co., 1939.

Bailey, Albert Edward. *Daily Life in Bible Times.* New York: Charles Scribner's Sons, 1943.

Bateman, John A. *Introduction to Highway Engineering.* London: John Wiley & Sons, 1942.

Bowen, John. *A Ship Modelmaker's Manual.* New York: Arco Publishing, 1982.

Brown, Curtis Maitland. *Evidence and Procedures of Boundary Location.* New York: John Wiley & Sons, 1962.

Bull, Robert J. "Caesarea Maritima: The Search for Herod's City." *Biblical Archaeology Review* (May/June 1982).

Canaan, Gershon. *Rebuilding the Land of Israel.* New York: Architectural Book Publishing Co., 1954.

Carta's Historic Atlas of Israel. Jerusalem: Carta, 1977.

Casson, Lionel. *The Ancient Mariners.* New York: Macmillan Co., 1959.

Central Intelligence Agency. *Issues of the Middle East.* Washington D.C.: U.S. Government Printing Office, 1973.

Chellis, Robert D. *Pile Foundations.* New York: McGraw-Hill Book Co., 1961.

Cornfeld, Gaalyah. *Archaeology of the Bible: Book by Book.* San Francisco: Harper & Row, 1976.

Cornfeld, Gaalyah. *Josephus—The Jewish War.* Grand Rapids, Mich.: Zondervan Publishing House, 1982.

Cucari, Artilio. *Ships—A History of Over 1000 Illustrations.* New York: Greenwich House, 1983.

De Camp, L. Sprague. *The Ancient Engineers.* Garden City, N.Y.: Doubleday & Co., 1963.

Derry, T. K. and William I. Trever. *A Short History of Technology.* New York: Oxford University Press, 1960.

Encyclopedia Judaica. Jerusalem: Keter Publishing House, 1978.

Ferrill, Arthur. *The Origins of War.* London: Thames and Hudson, 1985.

Fritz, Volkman. "Temple Architecture: What Can Archaeology Tell Us About Solomon's Temple." *Biblical Archaeology Review* (July/August 1987).

Gille, Bertrand. *Engineers of the Renaissance.* Cambridge, Mass.: M.I.T. Press, 1966.

Hamlin, A. D. *A Textbook of the History of Architecture.* London: Longwood Publishing Group, 1904.

Hodges, Henry. *Technology in the Ancient World.* New York: Alfred A. Knopf, 1974.

Housman, Erich and Edgar Slack. *Physics.* McGraw-Hill Book Co.

Institute of the History of Natural Sciences. *Ancient China's Technology and Science.* Beijing: Foreign Language Press, 1983.

Isaacson, Ben. *Dictionary of the Jewish Religion.* Englewood, N.J.: Bantam Books, 1979.

Keel, Othmar. *The Symbolism of the Biblical World.* New York: Seabury Press, 1978.

Kemp, Peter. *The History of Ships.* London: Orbis Publishing, 1978.

Kenyon, Kathleen M. *Jerusalem, Excavating 3000 Years of History.* London: Thames and Hudson, 1967.

Kitcher, Arthur. *The New Jerusalem.* London: Thames and Hudson, 1973.

Kollek, Teddy and Moshe Pearlman. *Jerusalem, Sacred City of Mankind.* Jerusalem: Steinmetzky's Agency, 1974.

Kutcher, Arthur. *The New Jerusalem—Planning and Politics.* London: Thames and Hudson, 1973.

Lanciani, Rodolfo Amedeo. *Ancient Rome in Light of Recent Discoveries.* Boston: Mifflin Co., 1888.

Landels, J. G. *Engineering in the Ancient World.* Los Angeles: University of California Press, 1981.

La Perrousaz, Ernest-Marie. "King Solomon's Wall Still Supports the Temple Mount." *Biblical Archaeology Review* (May/June 1987).

Lockyer, Herbert. *All the Trades and Occupations of the Bible: A Facinating Story of Ancient Arts and Crafts.* Grand Rapids, Mich.: Zondervan Publishing House, 1969.

Mansir, A. Richard. *A Modeler's Guide to Ancient and Medieval Ships to 1650.* Dana Point, Calif.: Moonraker Publication, 1981.

Mazar, Benjamin. *The Mountain of the Lord.* Tel Aviv: Hamikra Publications, 1975.

Merriman, Thaddeus. *American Civil Engineer's Handbook.* London: John Wiley & Sons, 1947.

Miller, John Anderson. *Atoms and Epochs.* Boston: Twayne Publishers, 1966.

Miller, Madelaine S. and J. Lane. *Harper's Encyclopedia of Bible Life.* New York: Harper & Row, 1978.

Morgan, Morris Hicky, ed. *Vitruvius—The Ten Books of Architecture.* New York: Dover Publications, 1960.

Muhly, James D. "How Iron Technology Changed the World and Gave the Philistines a Military Edge." *Biblical Archaeology Review* (November/December 1982).

Negev, Avraham. *Archaeological Encyclopedia of the Holy Land.* Englewood, N.J.: SBS Publishing, 1980.

Netzer, Ehud, "Jewish Rebels Dig Strategic Tunnel System." *Biblical Archaeology Review* (July/August 1988).

Paul, Shalom and William G. Dever. *Biblical Archaeology.* Jerusalem: Keter Publishing House, 1973.

"A Place of Pilgrimage," *Jerusalem Post* (January 3, 1987).

Rabinovich, Abraham. "1900th Anniversary of the Destruction." *Israel Magazine* (March 1971).

Ramsey, Charles George and Harold Reeve Sleeper. *Architectural Graphic Standards,* 5th ed. New York: John Wiley & Sons, 1961.

Reader's Digest Editors. *Atlas of the Bible—An Illustrated Guide to the Holy Land.* Pleasantville, N.Y.: Reader's Digest Assoc., 1981.

Reader's Digest Editors. *Great People of the Bible.* Pleasantville, N.Y.: Reader's Digest Assoc., 1981.

Reader's Digest Editors. *Quest for the Past.* Pleasantville, N.Y.: Reader's Digest Assoc., 1984.

Reich, Hanns. *Jerusalem.* New York: Hill and Wang, 1969.

Reid, Richard. *The Book of Buildings: Ancient, Medieval, Renaissance and Modern Architecture of North America and Europe.* New York: Van Nostrand Reinhold Co., 1983.

Rossnagel, W. E. *Handbook of Rigging.* 3rd ed. New York: McGraw-Hill Book Co., 1964.

Sed-Rajna, Gabrielle. *Ancient Jewish Art.* Secaucus, N. J.: Chartwell Books, 1985.

Siegner, Otto. *This Is Greece.* Munich, Germany: Ludwid Simon, 1955.

Silberman, Neil Asher. *Digging for God and Country.* New York: Alfred A. Knopf, 1982.

Straub, Hans. *A History of Civil Engineering.* Cambridge, Mass.: M.I.T. Press, 1962.

Ussiskkin, David. "Restoring the Great Gate at Lachish." *Biblical Archaeology Review* (March/April 1988).

Vogel, Steven. "Subtlety and Suppleness, Recurring Themes in the Mechanical Arrangement of Organisms." *Mechanical Engineering* (November 1986).

Wheeler, Mortimer. *Splendours of the East.* London: Spring Books, 1970.

Whiston, William. *Josephus' Complete Works.* Grand Rapids, Mich.: Kregel Publications, 1981.

Whiston, William. *The Works of Flavius Josephus—The Wars of the Jews.* Grand Rapids, Mich.: Baker Book House, 1979.

Winter, F. E. *Greek Fortifications.* Toronto: University of Toronto Press, 1971.

World Encyclopedia. Chicago: Field Enterprises Educational Corporation, 1962.

Yadin, Yigdal. *The Art of Warfare in Biblical Lands in the Light of Archaeological Study.* Vols. 1 and 2. New York: McGraw-Hill Book Co., 1963.

Yaggy, L. W. and T. L. Haines. *Museum of Antiquity: A Description of Ancient Life.* Chicago: Western Publishing House, 1881.

Index

Scripture Index